Optical Brain–Computer Interface

In this shortform book, Sun, French, and Unnithan explore state-of-the-art optical recording techniques, with a focus on the revolutionary miniaturized fluorescence microscope – the miniscope – for real-time and *in vivo* monitoring of multi-neuronal dynamics during cognition-related events.

The miniscope is a powerful tool that allows real-time *in vivo* optical recording of multi-neuronal activity in freely moving animals. This book highlights the use of the miniscope in the context of the hippocampus, a brain region crucial for memory and cognition. The authors employ a combination of theoretical concepts, practical applications, and illustrative case studies to deliver a comprehensive understanding of optical recording techniques. They provide step-by-step guides for using the miniscope, offer insights into data analysis, and discuss its implications in the context of hippocampal research and brain–computer interfaces. Readers will gain profound insights into the role of the hippocampus in memory and cognition, and expert knowledge of the latest miniaturized *in vivo* optical recording techniques. The book provides them with thorough guidance on implementing a miniaturized fluorescence microscope for a brain–computer interface and information on advanced analysis techniques on the activity of large neuronal populations.

This book provides an invaluable short and accessible guide for researchers, neuroscientists, and brain–computer interface enthusiasts to enable them to understand and leverage the immense potential of this advanced optical recording methodology.

Optical Brain–Computer Interface

Using a Miniscope to Detect Multi-Neuronal Dynamics during Cognition-Related Events

Dechuan Sun, Chris French, and
Ranjith R Unnithan

CRC Press
Taylor & Francis Group
Boca Raton London New York

CRC Press is an imprint of the
Taylor & Francis Group, an **informa** business

First edition published 2025
by CRC Press
2385 NW Executive Center Drive, Suite 320, Boca Raton FL 33431

and by CRC Press
4 Park Square, Milton Park, Abingdon, Oxon, OX14 4RN

CRC Press is an imprint of Taylor & Francis Group, LLC

ISBN: 9781032746807 (hbk)
ISBN: 9781032746821 (pbk)
ISBN: 9781003470397 (ebk)

DOI: 10.1201/9781003470397

Typeset in Times
by Newgen Publishing UK

Contents

Preface   vii
Acknowledgments   ix
About the Authors   xi

1 General Overview and Introduction   **1**
  1.1   General Introduction   1
  1.2   The Functions of the Hippocampus   3
  1.3   Multisensory Integration of Hippocampal Neural
        Network   4
  1.4   Brain Signal Recording Techniques   5
  1.5   Miniaturized Fluorescence Microscope   7
  1.6   Real-Time Brain–Computer Interface   8

2 Optical Manipulation Procedures   **14**
  2.1   Introduction   14
  2.2   Stereotaxic Surgery   15

**3 Scopolamine Impairs Spatial Information Recorded with
    "Miniscope" Calcium Imaging in Hippocampal Place Cells**   **17**
  3.1   Introduction   17
  3.2   Materials and Methods   18
  3.3   Results   24
  3.4   Discussion   31

**4 Hippocampal Cognitive and Relational Map Paradigms
    Explored by Multisensory Encoding Recording with
    Widefield Calcium Imaging**   **38**
  4.1   Introduction   38
  4.2   Results   41
  4.3   Discussion   54
  4.4   Methods   59

5   Real-Time Multimodal Sensory Detection Using Widefield
    Hippocampal Calcium Imaging   **73**
    5.1   Introduction   73
    5.2   Results   74
    5.3   Discussion   84
    5.4   Methods   88

6   Conclusions and Future Work   **99**
    6.1   Key Findings   99
    6.2   Future Work   100

Index   103

Preface

In recent years, technological advancements have significantly enhanced the field of neuroscience, enabling researchers to gain unprecedented insights into brain dynamics during cognition-related events. The miniaturized fluorescence microscope, or miniscope, has emerged as a powerful brain–computer interface, allowing *in vivo* optical recording of multi-neuronal activity in freely moving animals. This book aims to highlight the use of such advancements, particularly in the context of the hippocampus, a brain region crucial for memory and cognition. A noteworthy development is the recent attempt to develop high channel count brain–computer interfaces using flexible electrodes. This illustrates the immense interest and potential of these devices, addressing the critical issue of low channel count. Similarly, the miniaturized optical recording technique supports an alternative method for recording very large neuronal ensembles, and this cutting-edge technology holds promise as a superior tool in recording complex neural dynamics.

This book serves as a valuable resource for researchers and neuroscientists, providing insights and guidance on the use of miniaturized fluorescence microscope recording techniques, as well as related advanced analysis methods. After reading this book, the readers will be able to get :

- Profound insights into the role of the hippocampus in memory and cognition.
- Expert knowledge of the latest miniaturized *in vivo* optical recording techniques.
- Thorough guidance on implementing a miniaturized fluorescence microscope for a brain–computer interface.
- Advanced analysis techniques on the activity of large neuronal populations.

Dechuan Sun
April, 2024

Acknowledgments

I would like to express my gratitude to the following individuals for their invaluable input at various stages of this book's development or for their professional comments on specific chapters:

Daniel Aharoni (University of California, Los Angeles)
David Grayden (The University of Melbourne)
Anthony Burkitt (The University of Melbourne)
Yang Yu (The University of Melbourne)
Noor E Karishma Shaik (The University of Melbourne)
Forough Habibollahi Saatlou (The University of Melbourne)

I wish to extend my sincere appreciation to my coauthors, Ranjith R Unnithan and Chris French, for their invaluable contributions to the preparation and writing of this book.

I thank my publisher Andrew Stow and Kasturi Ghosh, for their dedication and exuberance.

Dechuan Sun

About the Authors

Dechuan Sun is a postdoctoral research fellow in the Department of Electrical and Electronic Engineering at The University of Melbourne.

Chris French leads the Neural Dynamics Laboratory at the Melbourne Brain Centre, is an Associate Professor in the Department of Medicine at The University of Melbourne and neurologist at the Royal Melbourne Hospital. He attained his PhD from the University of New South Wales and MD from the University of Sydney. He is a fellow of the Royal Australasian College of Physicians and a member of the Board of Directors of the Organisation for Computational Neuroscience (OCNS).

Ranjith R Unnithan is an associate professor in the Department of Electrical and Electronic Engineering at The University of Melbourne. He received his PhD in Electrical Engineering from the University of Cambridge. He is a member of IEEE, The International Society for Optics and Photonics, and the Australian Optical Society.

General Overview and Introduction

1

1.1 GENERAL INTRODUCTION

Cognition is characterized by intricate neural dynamics, whereby numerous neurons engage in interaction and coordination to process information, shape perceptions, make decisions, and execute actions. These multi-neuronal dynamics that underlie cognitive functions span various levels of organization, ranging from the individual neuron to interconnected neural networks. By exploring the complex neural networks and mechanisms underlying cognitive processes, researchers will gain a better understanding of brain function and get invaluable insights into the origins and progression of diverse neurological disorders. This enhanced understanding will serve as a foundation for devising more efficacious and precise therapeutic interventions for brain diseases.

To facilitate the study of brain activities for cognition research, the application of a brain–computer interface (BCI) is crucial. BCI technology offers two primary approaches: real-time BCI and non-real-time BCI, each with its own advantages and applications. Real-time BCIs process brain signals and provide feedback in real-time, typically within milliseconds, aiming to decode the brain signal in a fast way. In contrast, non-real-time BCIs process brain signals over longer periods and do not require immediate feedback. These systems are often used for applications such as data analysis, offline training, or research purposes where the timing of the interaction is not critical. Non-real-time BCIs can involve more complex and advanced signal processing and analysis techniques since there is no requirement for immediate response. For non-real-time BCIs, a synchronized camera or sensor is often utilized to track the concurrent events or behaviours of the subjects, which are then used for later analysis.

DOI:10.1201/9781003470397-1

High-resolution recording of brain cells has been possible since the early 20th century, but it was only recently that groups of neurons could be recorded, coinciding with the realization that such pattern changes (modulation) of neuronal ensembles were the basis for information encoding[1,2]. In terms of BCI recording techniques aimed at single-neuron resolution, there are typically two types. The first type is based on electroencephalography (EEG), which utilizes invasive electrode arrays to measure the electrical activity of small groups of neurons. The second type comprises advanced imaging techniques such as voltage-sensitive dye imaging and calcium imaging. Although significant improvements in invasive electrode technologies have occurred, the essential problem of inadequate numbers of recorded neurons and signal loss over time remain critical limitations[3]. For example, Lebedev, a pioneer in the field, reports that individual simple variables such as arm velocity can be decoded from 40 to 60 neurons, but that for more complex tasks and cognition-related events, a greater number of neurons would be needed. In comparison, the optical imaging method can provide a large field of view, monitoring the activity of a large number of neurons simultaneously for a long time, which is beneficial for neuronal population activity study.

This book contains six chapters and presents a comprehensive exploration of state-of-the-art optical recording techniques, with a focus on a revolutionary miniature fluorescence microscope known as the miniscope. This microscope functions as a BCI capable of both real-time and non-real-time applications. We utilize the hippocampus as an example to demonstrate its practical application. The aim of this book is to provide researchers, neuroscientists, and BCI enthusiasts with a valuable resource to understand and leverage the immense potential of this advanced optical recording methodology. The book employs a combination of theoretical concepts, practical applications to deliver a comprehensive understanding of this optical recording technique. It provides step-by-step guides for using the miniscope, offers insights into data analysis, and discusses its implications in the context of hippocampal research.

The first chapter begins with an introduction of the hippocampus, one of the most important brain regions involved in memory and cognition. We then present the development of *in vivo* brain signal recording techniques, and particularly highlight the advantages of the recently developed miniscope. Lastly, the chapter touches upon the fascinating realm of the real-time BCI. Overall, this chapter sets the stage for a compelling journey through the intricate world of the hippocampus and its relevance in optical recording and BCI research. The second chapter provides a detailed description of the optical manipulation procedure, showing the reader how to use the miniscope to record neuronal signals. The third chapter presents an illustrative case of utilizing the miniscope for an animal memory-related study. The study aims to characterize the properties of large neuronal populations and explore the mechanisms involved for information processing using a miniscope during the well-characterized cognitive impairment induced by scopolamine. The fourth chapter provides an example of using the miniscope for a cognition-related study, demonstrating the feasibility of employing the miniscope for detecting multisensory modalities in the hippocampus. The fifth chapter shows a real-time BCI application example of using the miniscope to decode hippocampal spatial and non-spatial information in real-time. The sixth chapter summarizes the work and points out some future development in this field.

After reading this book, the readers will be able to get: (1) profound insights into the role of the hippocampus in memory and cognition; (2) expert knowledge of the latest miniaturized *in vivo* optical recording techniques; (3) thorough guidance on implementing a miniaturized fluorescence microscope for a BCI; (4) advanced analysis techniques on the activity of large neuronal populations.

1.2 THE FUNCTIONS OF THE HIPPOCAMPUS

The hippocampus is a major component of the limbic system and plays an important role in memory and cognition[4-6]. Hippocampal neural ensembles are comprised of two major types of neurons – pyramidal cells and interneurons[7]. The pyramidal cells are primary excitation units in the hippocampus, which serve the function of transforming synaptic inputs to action potentials on the cellular level and controlling the transmission of hippocampal outputs to other brain regions. Previous work has proposed that excessive pyramidal cells excitation may induce neurological disorders, and death of these cells may lead to neuronal degeneration that is manifested as permanent impairments of cognition and[8,9]. In contrast, the hippocampal interneurons serve as modulatory elements to transmit information between local or remote brain regions and account for 10–15% of the neuronal population in the hippocampus. Currently, at least 21 different types of interneurons in the hippocampus have been identified, which modulate hippocampal neural activity in distinct ways[10]. The functions of interneurons are complex, but most of them are inhibitory ones that suppress and regulate the activity of local pyramidal cells, and dysfunction of interneurons may result in neuronal circuit disorders, such as epilepsy and speculatively schizophrenia[11].

Over the last century, a considerable amount of work has investigated the functional role of the hippocampus, and three main ideas dominate the area. The first idea is the inhibitory control of the neural circuit. This theory was originally proposed by Nadel in 1975[12]. Researchers observed hippocampal hyperactivity and enhanced neural oscillations in hippocampal lesioned animals. At that time, the understanding of the hippocampus was limited, and later, the effects were ascribed to the dysfunction of hippocampal inhibitory interneurons[11]. Nowadays, this hypothesis has gradually slipped out of popular interest. The second theory links the hippocampus to memory. This theory derived from the unexpected post-operative observations of a hippocampal lesion patient (Henry Molaison, also known as H.M.). Doctors found that this patient could not form any new episodic memories, but the memories formed many years ago were unaffected[13]. Episodic memory is a type of long-term memory that contains detailed contexts, such as past personal experiences, time, and location. Similar symptoms were later witnessed and reported in other patients with hippocampal damage or other species with hippocampal lesions, leading to the conclusion that successful episodic memory formation and retrieval required intact hippocampal circuitry[14-15]. The third idea relates the hippocampus to spatial cognition. In 1971, O'Keefe and Dostrovsky discovered a group of neurons in the rat hippocampus that was sensitive to the location of the animals, which were termed "place cells"[16]. Place cells are hippocampal pyramidal cells that fire when the animal comes into a particular space in a familiar environment, which helps the animal with spatial information processing. Since that time, many studies focused on the place sensitivity of hippocampal neurons, especially in cognitive task-related experiments[17-18]. Several studies found that the firing patterns of place cells were quite stable in a familiar environment even when some spatial cues were removed, indicating a strong involvement of place cells in both memory encoding and retrieval processes and a possible reconciliation between episodic memory and spatial cognition[18].

Many clinical studies have reported that patients with cognitive disorders are usually accompanied by different degrees of spatial cognition impairments[19-20]. In transgenetic mouse models of neurological disorders, the stability of place cells substantially degraded in a familiar environment, and the place cells could not form stable firing patterns in a new environment[21]. Additionally, the rhythmic oscillation of place cells was also affected, producing abnormal fast gamma oscillations and altered phase of theta oscillations[22-23]. Thus, the place cell can be potentially utilized as a functional cellular level biomarker of hippocampal dysfunction in cognitive disorders.

1.3 MULTISENSORY INTEGRATION OF HIPPOCAMPAL NEURAL NETWORK

The recent findings that rhythmic light flicker can rescue hippocampal low gamma oscillation and improve cognition in Alzheimer's disease mouse models[24-25] suggest an important role of non-spatial information in tuning hippocampal neural activity. Additionally, clinical studies in humans have found that the hippocampus plays critically important roles in episodic memory for a broad range of materials[26-27]. The hippocampus receives neural signals from the surrounding neocortex through the entorhinal cortex. As the hippocampus is in a complex set of neural loops receiving and processing multisensory signals, it also likely shows a certain degree of sensitivity to non-spatial information in addition to spatial context.

In clinical studies, patients with hippocampal damage usually have difficulty in completing non-spatial cues (such as images, words, and voices) guided tasks. Additionally, many have observed strong hippocampal activation during both spatial and non-spatial memory retrieval processes (see Eichenbaum, 2016 for review) [27]. In contrast, the picture is less clear in animal studies. In O'Keefe and Nadel's (1978) review[28], many studies did not find any significant effects of hippocampal damage on discrimination and conditioning learning experiments. Moreover, hippocampal lesions did not impact the animal's performance in visual or auditory cue-guided experiments[29-31]. However, non-spatial stimuli evoked hippocampal responses have also been reported, such as flash-light[32-33], artificial vowels at different centre frequencies[32,34,35], odours[36-38], and gustatory stimuli[39]. However, in these studies, the stimuli were given in task-related experiments, raising the possibility that the stimuli-evoked neural responses were position-specific. Additionally, most of these studies only macroscopically analysed the stimuli evoked neuronal responses by averaging the data recorded over trials, detecting sensitive neurons by measuring the information content, mutual information, or using a shuffling method. Hence, effects that might be related to ensemble neural dynamics, a fundamental functional level of neural information processing, have not been explored. It has been hypothesized that neuronal ensembles rather than a single neuron may encode neural signals more comprehensively, in what is termed as "population coding". In population coding, the neural representation is determined by the neural firing patterns of neuronal ensembles instead of just an individual cell[40]. For example, the position of an animal in place fields can be more accurately estimated by analysing the joint activities of place cell ensembles rather than induvial units[23,41,42]. However, whether the non-spatial information can be decoded in the same routine has not been investigated before.

1.4 BRAIN SIGNAL RECORDING TECHNIQUES

Although animal behavioural experiments provide intuitive and comprehensible and quantifiable aspects of cognition, it is critically important to investigate the dynamics of neuronal ensembles to understand these processes at a fundamental level. There are several commonly used recording techniques to study neural activity in animals' brains. EEG is a commonly used non-invasive method to record electrophysiological activity of the brain. Surface electrodes are fixed on the scalp to measure the electrical activity of the brain structures underneath, and each electrode receives a combination of signals coming from different brain structures with varying attenuations. While the direct signals from spatially arranged electrodes can be very informative about states of arousal or abnormal brain states such as encephalopathies, more sophisticated analyses can be performed. Principal component analysis (PCA), independent component analysis algorithms, or more advanced machine learning techniques can be applied to decompose the signals[43]. The main source of the EEG signal is the postsynaptic potentials of cortical neurons[44], neuronal action potentials cannot be accurately recorded, so this technique can only show the macroscopic activity of the brain with poor spatial resolution.

Another recording method is the use of intracranial electrodes, which is the most commonly used method in animal studies. Compared with EEG, this technique provides more precise recording with a much better signal-to-noise ratio. The electrodes are directly implanted into the regions of interest to record localized signals. Depending on the purpose of the study, there are several different configurations for these electrodes. For example, a single electrode with a diameter of 100–300 µm is commonly used to record the local field potentials; tetrodes (four twisted wires) or multichannel silicon probe with a diameter of 2–15 µm tips are widely used to record action potentials. After recording, a spike sorting algorithm[45] can be applied to extract the spike trains of each detected neuron.

A third recording method that has become popular in recent years is *in vivo* calcium imaging. This method takes advantage of fluorescent calcium indicators and optically measures neuronal calcium activity. When neurons fire, an influx of extracellular calcium occurs allowing action potentials to be detected in real time not only from individual neurons but also from functional groups of neurons, typically hundreds of cells. The signal generated is quite large, approximately 30% change in fluorescence per 100 mV. In the presence of calcium sensitive fluorescent indicators, the neuronal calcium activity of large numbers of neurons can be tracked simultaneously. There are several ways of delivering calcium fluorophores, but a relatively new and highly effective method is to use "genetically encoded calcium indicators" or GECIs. These are generated in neuronal cell bodies by using adeno-associated viral (AAV) transfection of DNA coding for the GECI of choice, typically one of the GCaMP family. Notably, these same AAV vectors are being used extensively for gene therapy delivery in many human clinical trials, including central nervous system delivery. The selection of calcium indicators depends on the purpose of the study. For example, GCaMP6s or jGCaMP7s is more sensitive to small changes in intracellular calcium levels and can show brighter neuronal signals. In contrast, GCaMP6f or jGCaMP7f is a fast variant that has a shorter response time and can track more precise temporal firing patterns. Compared with electrical signal recording methods, this method can provide the highest spatial resolution and a large field of view that contains many neurons. However, a disadvantage of this method is the slow calcium kinetics. Even the fastest variant GCaMP6f with a temporal resolution of about 50 ms, cannot accurately track the action potentials if the neuronal firing rates are too high[46].

1.5 MINIATURIZED FLUORESCENCE MICROSCOPE

Previous methods for real-time functional multi-neuronal microscopy in behaving mice require tabletop optical instrumentations, and the animals are usually head fixed on a running wheel to encourage movement[47]. This cannot be readily distributed to many mice in parallel and the fixation is incompatible with standard rodent behavioural assays. Additionally, these techniques are limited to relatively short, often terminal observation periods. The development of an open-source miniaturized fluorescence microscope (miniscope) has become widely adapted in the past 5 years[48]. Several labs have released their versions of miniscopes with versatile functions, such as UCLA miniscope, NINscope, Finchscope, CHEndoscope, and Inscopix miniscope, but the functional designs are similar. Among them, UCLA miniscopes (Figure 1.1a) have achieved broad dissemination and may be the most impactful due to the detailed online documentation and surgical tutorials as well as the analysis pipelines. The miniscope has a mass of around 3 g including a CMOS image sensor and can image an area of about 0.5 mm^2 in the brain. The system contains a lightweight body structure, optical lenses and filters, CMOS sensor, and readout electronics. A flexible coaxial cable is used to carry image data, control signals, and power from a data acquisition board. The functional design of miniscope is very similar to a traditional benchtop fluorescence microscope, but only keeping the most essential components: (i) The objective lens is replaced with a gradient refractive index (GRIN) lens to focus and relay the optical signal; (ii) a surface mount LED provides the stimulation light source; (iii) excitation filter, dichroic mirror, and emission filter are cut into small custom sizes; (vi) a small CMOS sensor with good quantum efficiency is placed on the top to collect emission light rather than a bulky CCD; (v) focal

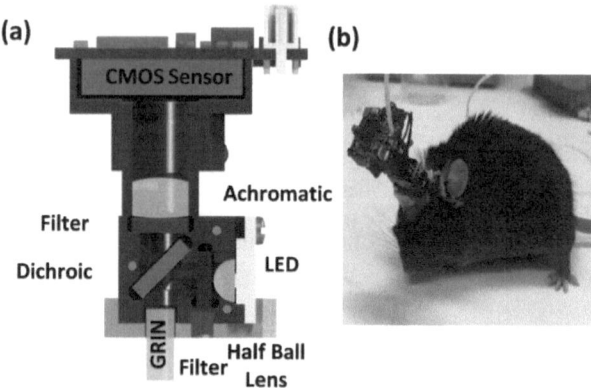

FIGURE 1.1 The structure of the miniscope. (a) Illustration of miniscope in situ – GRIN lens (blue cylinder) penetrates skull to image neurons in hippocampal CA1 with epifluorescence elements shown. (b) An example of the mouse with the miniscope.

length of the system is adjusted by moving the CMOS sensor instead of the objective lens. In the latest version published at the end of 2020, a liquid lens is used to automatically adjust the focal length and an inertial measurement unit is added to extend features. A growing body of researchers has implemented miniscopes, providing a very versatile device for studying neural activity in the brain, especially in the hippocampus of freely moving mice[47-48] (Figure 1.1b). The miniscope works as a general fluorescence microscope. In a proper settled miniscope, the depth of focus is "0", which means the image of the neuron on the bottom of the GRIN lens focuses on the CMOS sensor. But the depth of focus can be changed manually from 0 to 325 μm. If deep imaging is needed, a longer GRIN lens (relay lens) can be implanted into the animal's brain.

1.6 REAL-TIME BRAIN–COMPUTER INTERFACE

A wide range of common neurological disorders such as stroke and trauma lead to devastating impairment of physical function and communication, impacting an estimated 20,000 Australians annually and millions of persons internationally are similarly impaired. Remarkably, brain signals can be used to actuate prosthetic and communication devices in paralysed patients. It has been possible for about 40 years to use brain-recorded signals to control devices providing functional prostheses for paralysed persons[49-50]. The most capable devices record from single neurons, by inserting fine wires into the brain together with a spike sorting decoding algorithm, thus forming a "brain–computer interface" or BCI. Limitations of this technique are that relatively small numbers of neurons can be recorded and that the signal tends to drop-off over time. A well-known example is the "Utah Array" providing 64 channels and above, capable of actuating robotic arms, but losing signal often within 6 months[51]. A noteworthy development is the recent attempt to develop high channel count BCIs using flexible electrodes by Elon Musk[52] – this illustrates the immense interest and potential of these devices, addressing the critical issue of low channel count, but Musk's technique is extremely complex and there is no clear evidence that the signal will not attenuate even with the highly flexible electrodes used.

Due to the safety reliability, and long-term effectiveness issues, most BCI devices, are still tested on animals, and there is a huge gap before moving BCI devices from the laboratory to human end users. Although the target site for BCIs has generally been cortical areas specifically related to volitional movement, such as the motor cortex, other areas such as the hippocampus have also been found to be effective recording targets[41,53-55]. These experiments either use a single electrode or electrode arrays to record and analyse neuronal electrical signals but are still restricted to limited recording channels. In contrast, the calcium imaging methodology allows the observation of much greater numbers of neurons than conventional methods. This will greatly functionally enhance this interface as more neurons are recorded, and permit more complex analysis tasks. Additionally, the optical signals are quite stable even after a long period of time[56-57], which is extremely suitable for long-term experiments and potential therapeutic devices. Regarding the utility of calcium signals for BCI use, the miniscope has not been used yet for BCIs but two-photon calcium imaging of only 10 layer II/III neurons in mouse motor cortex has allowed volitional control of an external device[58], demonstrating clear proof of principle for the proposed methodology. Miniscopes thus hold great promise for BCIs, and the development of hippocampal BCIs will facilitate future closed-loop experiments.

REFERENCES

1. Lukashin, A.V., Amirikian, B.R. and Georgopoulos, A.P., 1996. A simulated actuator driven by motor cortical signals. *Neuroreport*, 7(15), pp.2597–2602.

2. Bouton, C.E., 2018. Advances in invasive brain–computer Interface technology and decoding methods for restoring movement and future applications. In *Neuromodulation* (pp. 415–425). Cambridge, MA: Academic Press.

3. Shoffstall, A.J. and Capadona, J.R., 2018. Bioinspired materials and systems for neural interfacing. *Current Opinion in Biomedical Engineering*, 6, pp.110–119.

4. Burgess, N., Maguire, E.A. and O'Keefe, J., 2002. The human hippocampus and spatial and episodic memory. *Neuron*, 35(4), pp.625–641.

5. Bird, C.M. and Burgess, N., 2008. The hippocampus and memory: insights from spatial processing. *Nature Reviews Neuroscience*, 9(3), pp.182–194.

6. Voss, J.L., Bridge, D.J., Cohen, N.J. and Walker, J.A., 2017. A closer look at the hippocampus and memory. *Trends in Cognitive Sciences*, 21(8), pp.577–588.

7. English, D.F., McKenzie, S., Evans, T., Kim, K., Yoon, E. and Buzsáki, G., 2017. Pyramidal cell-interneuron circuit architecture and dynamics in hippocampal networks. *Neuron*, 96(2), pp.505–520.

8. Sweatt, J.D., 2004. Hippocampal function in cognition. *Psychopharmacology*, 174(1), pp.99–110.

9. Klausberger, T. and Somogyi, P., 2008. Neuronal diversity and temporal dynamics: the unity of hippocampal circuit operations. *Science*, 321(5885), pp.53–57.

10. Deng, X., Gu, L., Sui, N., Guo, J. and Liang, J., 2019. Parvalbumin interneuron in the ventral hippocampus functions as a discriminator in social memory. *Proceedings of the National Academy of Sciences*, 116(33), pp.16583–16592.

11. Pelkey, K.A., Chittajallu, R., Craig, M.T., Tricoire, L., Wester, J.C. and McBain, C.J., 2017. Hippocampal GABAergic inhibitory interneurons. *Physiological Reviews*, 97(4), pp.1619–1747.

12. Nadel L, O'Keefe J, Black A (Jun 1975). "Slam on the brakes: a critique of Altman, Brunner, and Bayer's response-inhibition model of hippocampal function". *Behavioral Biology*, 14(2), pp.151–162.

13. Scoville, W.B. and Milner, B., 1957. Loss of recent memory after bilateral hippocampal lesions. *Journal of Neurology, Neurosurgery, and Psychiatry*, 20(1), p.11.

14. Squire, L.R., 1992. Memory and the hippocampus: a synthesis from findings with rats, monkeys, and humans. *Psychological Review*, 99(2), p.195.

15. Eichenbaum, H., 1993. *Memory, Amnesia, and the Hippocampal System*. Cambridge, MA: MIT Press.

16. O'Keefe, J. and Dostrovsky, J., 1971. The hippocampus as a spatial map: preliminary evidence from unit activity in the freely-moving rat. *Brain Research*, 34(1), pp.171–175.

17. Best, P.J., White, A.M. and Minai, A., 2001. Spatial processing in the brain: the activity of hippocampal place cells. *Annual Review of Neuroscience*, 24(1), pp.459–486.

18. Nakazawa, K., McHugh, T.J., Wilson, M.A. and Tonegawa, S., 2004. NMDA receptors, place cells and hippocampal spatial memory. *Nature Reviews Neuroscience*, 5(5), pp.361–372.

19. Khachaturian, Z.S., 1985. Diagnosis of Alzheimer's disease. *Archives of Neurology*, 42(11), pp.1097–1105.

20. Goedert, M. and Spillantini, M.G., 2006. A century of Alzheimer's disease. *Science*, 314(5800), pp.777–781.
21. Cacucci, F., Yi, M., Wills, T.J., Chapman, P. and O'Keefe, J., 2008. Place cell firing correlates with memory deficits and amyloid plaque burden in Tg2576 Alzheimer mouse model. *Proceedings of the National Academy of Sciences*, 105(22), pp.7863–7868.
22. Mably, A.J. and Colgin, L.L., 2018. Gamma oscillations in cognitive disorders. *Current Opinion in Neurobiology*, 52, pp.182–187.
23. Shuman, T., Aharoni, D., Cai, D.J., Lee, C.R., Chavlis, S., Page-Harley, L., Vetere, L.M., Feng, Y., Yang, C.Y., Mollinedo-Gajate, I. and Chen, L., 2020. Breakdown of spatial coding and interneuron synchronization in epileptic mice. *Nature Neuroscience*, 23(2), pp.229–238.
24. Martorell, A.J., Paulson, A.L., Suk, H.J., Abdurrob, F., Drummond, G.T., Guan, W., Young, J.Z., Kim, D.N.W., Kritskiy, O., Barker, S.J. and Mangena, V., 2019. Multi-sensory gamma stimulation ameliorates Alzheimer's-associated pathology and improves cognition. *Cell*, 177(2), pp.256–271.
25. Zheng, L., Yu, M., Lin, R., Wang, Y., Zhuo, Z., Cheng, N., Wang, M., Tang, Y., Wang, L. and Hou, S.T., 2020. Rhythmic light flicker rescues hippocampal low gamma and protects ischemic neurons by enhancing presynaptic plasticity. *Nature Communications*, 11(1), pp.1–16.
26. Wood, E.R., Dudchenko, P.A. and Eichenbaum, H., 1999. The global record of memory in hippocampal neuronal activity. *Nature*, 397(6720), pp.613–616.
27. Eichenbaum, H., 2016. What versus where: non-spatial aspects of memory representation by the hippocampus. *Behavioral Neuroscience of Learning and Memory*, pp.101–117.
28. O'keefe, J. and Nadel, L., 1978. *The Hippocampus as a Cognitive Map*. Oxford: Clarendon Press.
29. Morris, R.G., Garrud, P., Rawlins, J.A. and O'Keefe, J., 1982. Place navigation impaired in rats with hippocampal lesions. *Nature*, 297(5868), pp.681–683.
30. Phillips, R.G. and LeDoux, J.E., 1992. Differential contribution of amygdala and hippocampus to cued and contextual fear conditioning. *Behavioral Neuroscience*, 106(2), p.274.
31. Mumby, D.G., 2001. Perspectives on object-recognition memory following hippocampal damage: lessons from studies in rats. *Behavioural Brain Research*, 127(1–2), pp.159–181.
32. Sakurai, Y., 1996. Hippocampal and neocortical cell assemblies encode memory processes for different types of stimuli in the rat. *Journal of Neuroscience*, 16(8), pp.2809–2819.
33. Liu, Y.Z., Wang, Y., Tang, W., Zhu, J.Y. and Wang, Z., 2018. NMDA receptor-gated visual responses in hippocampal CA1 neurons. *The Journal of Physiology*, 596(10), pp.1965–1979.
34. Itskov, P.M., Vinnik, E., Honey, C., Schnupp, J. and Diamond, M.E., 2012. Sound sensitivity of neurons in rat hippocampus during performance of a sound-guided task. *Journal of Neurophysiology*, 107(7), pp.1822–1834.
35. Vinnik, E., Antopolskiy, S., Itskov, P.M. and Diamond, M.E., 2012. Auditory stimuli elicit hippocampal neuronal responses during sleep. *Frontiers in Systems Neuroscience*, 6, p.49.

36. Wiebe, S.P. and Stäubli, U.V., 1999. Dynamic filtering of recognition memory codes in the hippocampus. *Journal of Neuroscience*, 19(23), pp.10562–10574.
37. Wiebe, S.P. and Stäubli, U.V., 2001. Recognition memory correlates of hippocampal theta cells. *Journal of Neuroscience*, 21(11), pp.3955–3967.
38. Komorowski, R.W., Manns, J.R. and Eichenbaum, H., 2009. Robust conjunctive item–place coding by hippocampal neurons parallels learning what happens where. *Journal of Neuroscience*, 29(31), pp.9918–9929.
39. Ho, A.S., Hori, E., Thi Nguyen, P.H., Urakawa, S., Kondoh, T., Torii, K., Ono, T. and Nishijo, H., 2011. Hippocampal neuronal responses during signaled licking of gustatory stimuli in different contexts. *Hippocampus*, 21(5), pp.502–519.
40. Averbeck, B.B., Latham, P.E. and Pouget, A., 2006. Neural correlations, population coding and computation. *Nature Reviews Neuroscience*, 7(5), pp.358–366.
41. Guger, C., Gener, T., Pennartz, C., Brotons-Mas, J., Edlinger, G., Badia, B.I., Schaffelhofer, S., Verschure, P. and Sanchez-Vives, M.V., 2011. Real-time position reconstruction with hippocampal place cells. *Frontiers in Neuroscience*, 5, p.85.
42. Sun, D., Unnithan, R.R. and French, C., 2021. Scopolamine impairs spatial information recorded with "miniscope" calcium imaging in hippocampal place cells. *Frontiers in Neuroscience*, 15.
43. Subasi, A. and Gursoy, M.I., 2010. EEG signal classification using PCA, ICA, LDA and support vector machines. *Expert Systems with Applications*, 37(12), pp.8659–8666.
44. Kirschstein, T. and Köhling, R., 2009. What is the source of the EEG?. *Clinical EEG and Neuroscience*, 40(3), pp.146–149.
45. Rey, H.G., Pedreira, C. and Quiroga, R.Q., 2015. Past, present and future of spike sorting techniques. *Brain Research Bulletin*, 119, pp.106–117.
46. Chen, T.W., Wardill, T.J., Sun, Y., Pulver, S.R., Renninger, S.L., Baohan, A., Schreiter, E.R., Kerr, R.A., Orger, M.B., Jayaraman, V. and Looger, L.L., 2013. Ultrasensitive fluorescent proteins for imaging neuronal activity. *Nature*, 499(7458), pp.295–300.
47. Aharoni, D., Khakh, B.S., Silva, A.J. and Golshani, P., 2019. All the light that we can see: a new era in miniaturized microscopy. *Nature Methods*, 16(1), pp.11–13.
48. Aharoni, D. and Hoogland, T.M., 2019. Circuit investigations with open-source miniaturized microscopes: past, present and future. *Frontiers in Cellular Neuroscience*, 13, p.141.
49. Wolpaw, J.R., Birbaumer, N., Heetderks, W.J., McFarland, D.J., Peckham, P.H., Schalk, G., Donchin, E., Quatrano, L.A., Robinson, C.J. and Vaughan, T.M., 2000. Brain-computer interface technology: a review of the first international meeting. *IEEE Transactions on Rehabilitation Engineering*, 8(2), pp.164–173.
50. Vaughan, Theresa M., William J. Heetderks, Leonard J. Trejo, William Z. Rymer, Michael Weinrich, Melody M. Moore, Andrea Kübler et al. "Brain-computer interface technology: a review of the Second International Meeting." *IEEE Transactions on Neural Systems and Rehabilitation Engineering: A*

Publication of the IEEE Engineering in Medicine and Biology Society 11, no. 2 (2003): 94–109.

51. Maynard, E.M., Nordhausen, C.T. and Normann, R.A., 1997. The Utah intracortical electrode array: a recording structure for potential brain-computer interfaces. *Electroencephalography and Clinical Neurophysiology*, 102(3), pp.228–239.

52. Musk, E., 2019. An integrated brain-machine interface platform with thousands of channels. *Journal of Medical Internet Research*, 21(10), p.e16194.

53. Hoffmann, L.C., Cicchese, J.J. and Berry, S.D., 2015. Harnessing the power of theta: natural manipulations of cognitive performance during hippocampal theta-contingent eyeblink conditioning. *Frontiers in Systems Neuroscience*, 9, p.50.

54. Ciliberti, D., Michon, F. and Kloosterman, F., 2018. Real-time classification of experience-related ensemble spiking patterns for closed-loop applications. *Elife*, 7, p.e36275.

55. Hu, S., Ciliberti, D., Grosmark, A.D., Michon, F., Ji, D., Penagos, H., Buzsáki, G., Wilson, M.A., Kloosterman, F. and Chen, Z., 2018. Real-time readout of large-scale unsorted neural ensemble place codes. *Cell Reports*, 25(10), pp.2635–2642.

56. Ziv, Y., Burns, L.D., Cocker, E.D., Hamel, E.O., Ghosh, K.K., Kitch, L.J., El Gamal, A. and Schnitzer, M.J., 2013. Long-term dynamics of CA1 hippocampal place codes. *Nature Neuroscience*, 16(3), pp.264–266.

57. Rubin, A., Geva, N., Sheintuch, L. and Ziv, Y., 2015. Hippocampal ensemble dynamics timestamp events in long-term memory. *Elife*, 4, p.e12247.

58. Clancy, K.B., Koralek, A.C., Costa, R.M., Feldman, D.E., and Carmena, J.M. (2014). Volitional modulation of optically recorded calcium signals during neuroprosthetic learning. *Nature Neuroscience* 17, 807–809.

Optical Manipulation Procedures

2

2.1 INTRODUCTION

Over the past 5 years, there has been a notable increase in the number of open-source designs for miniaturized fluorescence microscopes (miniscopes) dedicated to medical, pharmacology, engineering, and neuroscience applications. This surge in development has not only propelled the progression of fluorescence microscope technologies but has also widened accessibility to a more expansive scientific community. Several laboratories have endeavoured to develop their own versions of the miniscope, making them available to the public domain. Among these endeavours, the UCLA miniscope stands out as arguably the most influential, given its extensive adoption globally.

While devices designed by different laboratories serve various functions, the working principles and manipulation procedures are quite similar. In this chapter, we use the UCLA miniscope version 3 as an example to demonstrate how to record brain activity in the hippocampus of rodents. Briefly, we first inject a virus into the animal's hippocampus to label neurons, causing them to exhibit fluorescent signals when exposed to stimulation light. After 1 week, a relay lens is implanted in the brain to collect these fluorescent signals. Approximately, 4 weeks later, we assess the signal quality using the miniscope and attach a baseplate to the animal's head, facilitating further recording.

DOI:10.1201/9781003470397-2

2.2 STEREOTAXIC SURGERY

The surgical procedures comprised three components: viral infusion, GRIN lens implantation, and baseplate mounting.

2.2.1 Virus Infusion

Sterilize all surgery instruments and get the animal ready for the surgery. At the start of the surgery, anaesthetize the mouse in a small induction chamber using an isoflurane induction rate of 3% and an oxygen flow of 1 L/min. After the animal is fully anaesthetized, fix the animal's head on a stereotaxic frame. Adjust the isoflurane flow rate to 2.5% and the oxygen flow to 0.25 L/min. Assess the depth of anaesthesia by conducting a toe-pinch test and wait until the animal no longer shows responsive behaviours. Shave the mouse's head using an electric shaver and sterilize the scalp using betadine and 80% ethanol. Apply eye ointment to ensure the protection of the eyes. Use a sharp blade to open the scalp and expose the skull. Align the skull using the landmarks Bregma and Lambda by tilting the animal's head to ensure they are at the same level. Utilize a drill burr to create a 0.4-mm hole at the coordinates of AP -2.1, ML +2.1. Then, inject the pAAV.Syn.GCaMP6f.WPRE.SV40 virus (500 nL, viral load: 2.2×10^{13} GC/mL; AddGene, USA) into the dorsal hippocampus through the drilled hole(AP -2.1, ML +2.1, DV -1.7 relative to bregma) with a custom made injecting system over 15 min. The virus is first loaded into a capillary (interior diameter: 0.9 mm), which is fabricated on a Sutter P-1000 electrode puller to produce a tip diameter of 20–50 μm; this is sealed with silicon oil (Sigma–Aldrich, USA) at the open end; and a round brass rod (diameter: 0.8 mm; Albion Alloys, UK), connected to 3D positioner is fitted into the capillary to control the volume of the virus injected. The virus injector is left in place for an additional 10 min to allow for viral diffusion. The animal is then left for 1 week to recover and to allow fluorophore expression.

2.2.2 GRIN Lens Implantation

On the day of the surgery, prepare the artificial cerebrospinal fluid (ACSF) solution using the following formula – NaCl: 127 mM; KCl: 1.2 mM; $CaCl_2$: 2.5 mM; $MgCl_2$: 1.5 mM; KH_2PO_4: 1.2 mM; D-glucose: 10 mM; $NaHCO_3$: 28 mM. Bubble the solution with carbogen (5% CO_2, 95% O_2) for at least 1 h.

Prepare the animal on the stereotaxic using the method mentioned above. Implant two 1-mm screws on animal's head (AP +1.8, ML -2.5; AP -2.8, ML -0.8) to serve as anchors. A small window of skull is removed by using a 2-mm drill bur, centred at AP -2.1, ML +1.6, and the exposed dura is cleaned with fine tweezers. A 27-gauge blunt needle is used to aspirate cortex to expose the vertical striations of the hippocampal fimbria, with ACSF flowing during the procedure to provide a clear operating field. Using the most posterior point of the edge of the drilled hole (next to lambda side) as a reference, the GRIN lens (0.25 pitch, #64-519, Edmund Optics) is implanted 1.35-mm deep, touching the surface of exposed tissue. Cyanoacrylate glue is applied surrounding the lens to prevent movement and dental cement is built over the glue for support. The lens is then covered with fast setting silicone adhesive (Dragon Skin® Series, USA). After the surgery, the animal is injected with carprofen (5 mg/kg) and dexamethasone (0.6 mg/kg, Sigma–Aldrich, USA) intraperitoneally every day for the relief of pain and inflammation and provided with enrofloxacin water (1:150 dilution, Baytril®, USA) for 1 week.

2.2.3 Baseplate Mounting and Recording

Four weeks later, prepare the animal and fix it on the stereotaxic arm. Carefully remove the silicone adhesive and clean the lens surface using alcohol. Connect the miniscope to the laptop thorough a DAQ board and open the control software. In the software, set the camera exposure and gain to the maximum, adjust the camera frame rate to 30 fps, set the stimulation LED intensity to proper values (typically around 4%), and adjust the focal length of the miniscope until clear neurons can be observed. After detecting clear signals, mount a small metal baseplate on the animal's head to support the miniscope, which is then locked into the optimal focal distance.

Scopolamine Impairs Spatial Information Recorded with "Miniscope" Calcium Imaging in Hippocampal Place Cells

3

3.1 INTRODUCTION

In recent years, optical brain–computer interfaces (OBCIs) have gradually been utilized in the field of medical research. In this chapter, we will describe a medical study that uses the OBCI to investigate the effects of the drug scopolamine on memory and provide details on how to analyse the recorded signals. In this study, a GRIN lens was implanted in the hippocampus, which serves as an OBCI.

DOI:10.1201/9781003470397-3

The formation and retrieval of episodic memory critically rely on the function of the hippocampus[1-2]. The spatial information processing ability of the hippocampus has been extensively studied, especially after the discovery of place cells by O'Keefe and Dostrovsky[3]. Place cells are hippocampal pyramidal neurons that activate in response to position and are believed to provide a spatial map of the environment[4-5]. Spatial memory in rodent hippocampus is affected by acetylcholine (ACh), with muscarinic acetylcholine receptors (mAChRs) playing a significant role[6]. Scopolamine is a non-specific antagonist of mAChRs, which has been commonly used in cognitive studies[7-8]. Blocking mAChRs with scopolamine has been found to greatly impair spatial memory encoding (see Klinkenberg and Blokland, 2010 for review)[9], but its effect on spatial memory retrieval is less clear. Some studies show no or very little influence of scopolamine on this (see Hasselmo, 2006 for review)[10], while others have reported impairment[11-12]. We hypothesized that wide field calcium imaging of the hippocampal CA1 region would allow a more detailed understanding of the effects of mAChR on the storage and retrieval of spatial information. Previous studies have used intracranial electrode arrays to record local field potentials and single-unit activity, allowing detailed observations of cognition-related processes in terms of neural firing patterns. However, the quantity of cells and spatial accuracy are generally limited. We have therefore used *in vivo* calcium imaging to record the activity of neural ensembles in the hippocampal CA1 region of freely running mice with a miniaturized microscope (miniscope)[13], allowing the simultaneous recording of calcium activity of a large population of neurons. We demonstrated that scopolamine greatly impaired the spatial accuracy of place cells with mice traversing a linear track, attributable to impaired stability of spatial representation revealed by hippocampal neural ensemble. Several parameters including neural firing rate, spatial information, total neuron number were affected. We found similar results in a conductance-based hippocampal network with scopolamine effects modelled by disinhibition of muscarinic modulation of voltage-gated potassium channels.

3.2 MATERIALS AND METHODS

All surgical and experimental procedures were approved by the Florey Animal Ethics Committee (No. 18-008UM) and were conducted in strict accordance with the Australian Animal Welfare Committee guidelines.

3.2.1 Subjects

Five naive adult male C57BL/6 mice aged 12 weeks were obtained from WEHI (Melbourne, VIC) and housed in the Biological Research Facility of the Department of Medicine, University of Melbourne. All animals weighed 24–25 g (24.66 ± 0.10 g) at the time of surgery and were housed individually. The facility was maintained on a 12–12h light–dark schedule (lights on: 7:30 am–7:30 pm) with water and standard mice chow *ad libitum*

3.2.2 Drug Administration

Scopolamine hydrobromide (Sigma–Aldrich, USA) was dissolved in sterile 0.9% saline and injected intraperitoneally at a volume of 1 mg/kg.

3.2.3 Virus Infusion

The pAAV.Syn.GCaMP6f.WPRE.SV40 virus (500 nL, viral load: 2.2×10^{13} GC/mL; AddGene, USA) was injected into the dorsal hippocampus (AP -2.1, ML +2.1, DV -1.7 relative to bregma) using a microinjection system over 15 min. The animal was then left for 1 week to recover and to allow fluorophore expression.

3.2.4 GRIN Lens Implantation

A small window of skull was removed by using a 2-mm drill bur, centred at AP -2.1, ML +1.6, and the exposed dura was cleaned with fine tweezers. A 27-gauge blunt needle was used to aspirate cortex to expose the vertical striations of the hippocampal fimbria. The GRIN lens (0.25 pitch, #64-519, Edmund Optics) was implanted 1.35-mm deep. Cyanoacrylate glue was applied surrounding the lens to prevent movement and dental cement was built over the glue for support. After the surgery, the animal was injected with carprofen (5 mg/kg) and dexamethasone (0.6 mg/kg, Sigma–Aldrich, USA) intraperitoneally every day for the relief of pain and inflammation, and provided with enrofloxacin water (1:150 dilution, Baytril®, USA) for 1 week. Four weeks later, a small metal baseplate was mounted on the animal's head to support the miniscope, which was locked in the position at the optimal focal distance.

3.2.5 Animal Training

After surgery, the animals were handled approximately 10 min twice a day in the daytime and weighed after each handling session for 5 days. A food restriction regimen was implemented to keep the animal at 85% of its original weight. The animal was then trained to run back and forth on a 1.6-m linear track for 2 weeks with clues painted on the walls for food reward while wearing the miniscope. On each training day, the mice performed 30 trials and a small food pellet was awarded once it could rapidly run through the track without wandering.

3.2.6 Experimental Procedures

All the recordings were performed during the daytime. The animal was brought into a silent recording room 30 min before the start of recording to acclimate to the surrounding environment. After mounting the miniscope, the animal was moved to the linear track to explore the space for 30 min freely. The linear track was cleaned with 80% ethanol to eliminate scent clues. Twelve running trials were recorded as the baseline control. Imaging frames were recorded with custom-made miniscope acquisition software, with a sampling rate of 30 FPS. The animal was then injected with saline or scopolamine, replaced in its cage for 20 min and then performed another 12 running trials. A camera fixed overhead was synchronized with the miniscope to record the animal's position. The miniscope cable was suspended over the linear track (Figure 3.1).

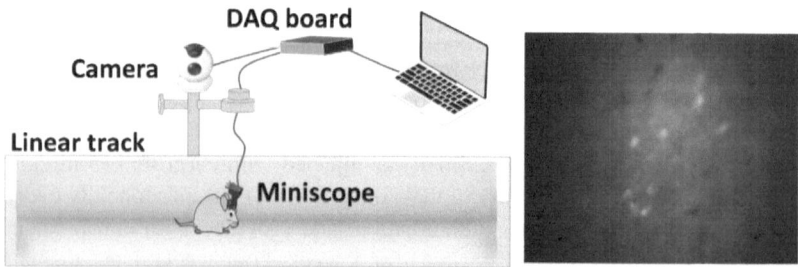

FIGURE 3.1 Linear track and miniscope recording system and an example of the raw image.

3.2.7 Image Pre-Processing and Calcium Activity Deconvolution

A non-rigid motion correction algorithm was applied first to implement image registration[14]. Constrained Non-negative Matrix Factorization for microendoscopic data were then utilized to identify and extract each neural spatial boundary and calcium activity[15]. A fast deconvolution algorithm was finally applied to deconvolve the calcium activity to estimate neural spike-activity[16].

3.2.8 Place Field Map

The position of the animal's head and running speed were detected using a custom Matlab script. We analysed the data of both left-to-right (LR) and right-to-left (RL) running directions separately. To analyse the neural spatial spike activity, the linear track was divided into several 2-cm spatial bins (the bins on each end were discarded). A speed range threshold was set between 8 and 25 cm/s. The neural temporal calcium event rate (the number of spikes in each bin) and the occupancy of the animal were counted and smoothed with a Gaussian smoothing kernel ($\sigma = 1.5$, size = 5). The place field map for each neuron was measured by dividing neural smoothed spatial spike activity by the smoothed bins occupancy, with the maximum value defined as the place field's position[17].

3.2.9 Spatial Information Content and Place Cells

The neural spatial information content (SIC) was defined as[18]:

$$I = \sum_{i=1}^{K} P_i \frac{\lambda_i}{\bar{\lambda}} \log_2 \frac{\lambda_i}{\bar{\lambda}} \tag{3.1}$$

K represents the number of bins; P_i is the occupancy ratio of the ith bin; λ_i is the neural calcium event rate in the ith bin; $\bar{\lambda}$ is the mean calcium event ($\sum_{i=1}^{K} P_i \lambda_i$).

We first calculated the SIC for each neuron, and then shuffled the animal's position as well as neural temporal spike activity 1000 times. A place cell was defined as the neuron whose SIC was above chance ($p < 0.05$) with respect to the shuffling results.

3.2.10 Odd and Even Trials Population Vector Overlap

To quantify the stability of spatial representation revealed by hippocampal neural ensemble, we calculated the population vector overlap (PVO) between odd and even trials[18]. The PVO is a measure of firing patterns' similarity across locations on the linear track. A high PVO value indicates similar firing patterns at two locations.

$$PVO(x,y) = \frac{\sum_{n=1}^{N} \lambda_n(x)\lambda_n(y)}{\sqrt[2]{\sum_{n=1}^{N} \lambda_n(x)\lambda_n(x)} \cdot \sqrt[2]{\sum_{n=1}^{N} \lambda_n(y)\lambda_n(y)}} \qquad (3.2)$$

N represents the total number of neurons; x and y represent the animal's location bin in odd and even trials, respectively; λ_n is the place field map of the nth neuron at different location bins.

The diagonal index (DI) is used to quantify the "diagonalization feature" of the PVO map. A larger value indicates more evident "diagonalization feature".

$$DI = \frac{1}{\sqrt{\sum_x \sum_y \left[PVO(x,y) - I \right]^2}} \qquad (3.3)$$

x and y represent the animal's location bin in odd and even trials respectively; I is an identity matrix (diagonal elements equal to 1).

3.2.11 Autoencoder

An autoencoder is a type of artificial neural network that learns how to compress and reconstruct data in an unsupervised way. It is a powerful tool for dimensionality reduction[19]. An autoencoder comprises an encoder and a decoder. We implemented and trained an autoencoder model *by* using the TensorFlow platform[20]. The encoder and decoder were both defined with three "Dense Layers" with "scaled exponential linear unit" (SELU) activation function (100 nodes in the last layer of encoder). The whole model was compiled with a binary cross-entropy loss function and Adam optimizer. We used a grid search method to fit the parameters including learning rate, epochs, and nodes for each animal to minimize the reconstructed data error. We wanted to explore if there were differences before and after mAChRs blockade that were reflected on the neural ensemble level. The calcium spike trains of the same neurons at baseline and saline/scopolamine were split into 2.5-s epochs and fed into the model. PCA of the encoder analysis was applied for better visualization.

3.2.12 Decoding

A naive Bayesian decoder was utilized to estimate the position of the animal based on neuronal temporal spike activity[21]. The decoding error was defined as the average distance between the real and estimated position at each time frame.

$$P(x|n) = P(x)\left(\prod_{i=1}^{N} f_i(x)^{n_i}\right) \cdot \exp\left(-\tau \sum_{i=1}^{N} f_i(x)\right) \tag{3.4}$$

The symbol n is the current input temporal spike activity; x is the index of the location bin; $P(x)$ is the occupancy ratio of the bin x; $f_i(x)$ is the average calcium event rate of the ith neuron at bin x; τ is the time window length of the input temporal spike activity.

3.2.13 Hippocampal CA1 Network Model

We sought to explore the experimental results in a conductance-based CA1 network derived from models that composed of 130 pyramidal neurons (PYR), 8 basket cells (BC), 2 axoaxonic cells (AAC), 2 bistratified cells (BIS), 2 oriens lacunosum-moleculare cells (OLM), 1 vasoactive intestinal peptide–cholecystokinin cell (VIP-CCK), and 4 vasoactive intestinal peptide–calretinin cells (VIP-CR) with the same neural characteristics and structures[22-23]. Additionally, we added an M-channel conductance on the models of somatostatin positive (SST+) and parvalbumin positive (PV+) interneurons with the conductance of 2 ps / m^2. The network connectivity is shown in Figure 3.2, and the afferent inputs to CA1 stored in CA3 were simulated by applying a spatial filter on the grid-like inputs from the entorhinal cortex (EC). We hypothesized that a major effect of scopolamine would be the enhancement of the muscarinic-inhibited M-current that is known to strongly modulate neuronal firing rates both in pyramidal neurons and interneurons (see "Discussion"). To simulate the effects of scopolamine, the M-channel conductances of pyramidal cells were increased by five (×5), ten (×10), fifteen (×15), and twenty (×20) times separately. The M-channel conductance amplitude of SST+ and PV+ interneurons were set at a level reproducing the neural firing reduction observed experimentally.

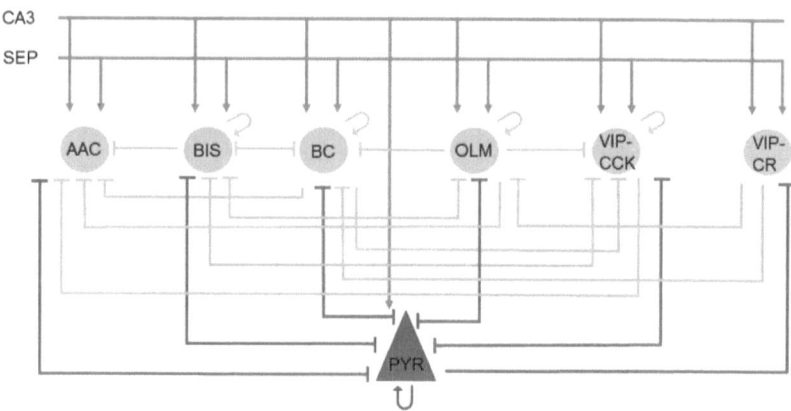

FIGURE 3.2 Schematic of the hippocampal network model. The model contains connectivity of pyramidal cells (PYR), basket cells (BC), axoaxonic cells (AAC), bistratified cells (BIS), oriens lacunosum-moleculare cells (OLM), vasoactive intestinal peptide-cholecystokinin cells (VIP-CCK), and vasoactive intestinal peptide–calretinin cells (VIP-CR). The afferent inputs come from hippocampal CA3 and medial septum (SEP).

3.2.14 Statistics

Statistical analyses were performed using SPSS (IBM) and GraphPad Prism software. The level of significance was set at $p < 0.05$. Statistical significance was assessed by two-tailed paired Student's t-test, one-way and two-way repeated-measures analysis of variance (ANOVA). Normality of the data was confirmed by Kolmogorov–Smirnov and Shapiro–Wilks test. Bonferroni correction was performed for the *post hoc* comparisons. No statistical methods were used to predetermine sample sizes, but our sample sizes were similar to those reported previously[23]. All results are shown as mean ± standard error of mean (SEM) of the percentage variation from baseline.

3.3 RESULTS

We compared the firing patterns of the place cell ensembles between control and treatment groups and measured total cell numbers, place cell numbers, neural firing rate, place field map, spatial information, odd and even trials PVO, decoding error, as well as the animal's running speed. All results are shown as a percentage change following injection of saline or scopolamine relative to baseline.

3.3.1 Scopolamine Reduced the Total Cell Number of Cells Detected and the Neural Firing Rate

We first studied the effects of blocking mAChRs on the number of detected neurons and the neural firing rate. Scopolamine significantly reduced the number of detected neurons (84.7% ± 9.10% relative to baseline) compared with saline (101.8% ± 1.45% relative to baseline; paired t-test, $p = 0.048$). Prior to the injection of scopolamine, there were 640 ± 46 neurons recorded in each animal, decreasing to 541 ± 46 following injection. Additionally, we found that the scopolamine reduced the neural firing rate to 83.39% ± 2.22% relative to baseline. This was significantly lower than saline controls with 98.72% ± 1.50% relative to baseline (paired t-test, $p = 0.006$). Prior to the injection of scopolamine, the average neural firing rate was 0.077 ± 0.012 Hz, and it decreased to 0.063 ± 0.009 Hz after the administration of scopolamine.

3.3.2 Reduction of Place Cell Number and Spatial Information Content Occurred with Scopolamine Administration

Head direction plays an important role in rodent spatial navigation, so we analysed the data separately on the running trials of RL and LR. Prior to the injection of scopolamine, 354 ± 26 place cells were observed in each animal in the RL and 323 ± 24 place cells in the LR running sessions. After the administration of scopolamine, the numbers reduced to 203 ± 32 (RL) and 203 ± 33 (LR). The percentage of place cells in both running directions significantly declined after scopolamine injection (RL: 57.63 ± 8.10% and LR: 62.73 ± 9.06% relative to baseline) compared with saline (RL: 109.43 ± 2.10% and LR: 100.49 ± 3.43% relative to baseline) (two-way repeated-measures ANOVA; main effect of injection, $F1,4 = 22.19$, $p = 0.009$; main effect of running direction, $F1,4 = 0.26$, $p = 0.637$; interaction, $F1,4 = 5.60$, $p = 0.077$; *post hoc* test saline versus scopolamine: RL, $p = 0.0099$; LR, $p = 0.0477$). SIC, a measure of location sensitivity of neurons was then quantified. The average SIC of all cells observed was significantly lower in scopolamine treated mice (RL: 74.35% ± 4.91% and LR: 75.14% ± 5.47% relative to baseline) than saline treated mice (RL: 102.44% ± 4.36% and LR: 99.00% ± 3.18% relative to baseline) (two-way repeated-measures ANOVA; main effect of injection, $F1,4 = 17.11$, $p = 0.014$; main effect of running direction, $F1,4 = 0.46$, $p = 0.535$; interaction, $F1,4 = 5.68$, $p = 0.076$; *post hoc* test saline versus scopolamine: RL, $p = 0.0249$; LR, $p = 0.0036$). The absolute SIC values were $1.71 ± 0.17$ bits and $1.63 ± 0.13$ bits on RL and LR running directions prior to scopolamine treatment, decreasing to $1.29 ± 0.20$ bits and $1.25 ± 0.18$ bits after injection. Additionally, the number of neurons with high SIC (>2 bits) was significantly reduced in the scopolamine group (RL: 22.37% ± 4.59% and LR: 29.51% ± 7.21% relative to baseline) compared with the control group (RL: 111.56% ± 12.02% and LR: 99.65% ± 6.93% relative to baseline) (two-way repeated-measures ANOVA; main effect of injection, $F1,4 = 30.89$, $p = 0.005$; main effect of running direction, $F1,4 = 0.28$, $p = 0.623$; interaction, $F1,4 = 23.58$, $p = 0.008$; *post hoc* test saline versus scopolamine: RL, $p < 0.0087$; LR, $p < 0.0133$).

3.3.3 Neuronal Ensemble Properties Were Impaired by Scopolamine

In addition to the individual neuron properties described above, we explored the effects of scopolamine on neural ensemble activity by calculating the animal's place field map (see "Methods", Figure 3.3). Normally, the firing of place cells displayed very high stability within the place fields. Before injecting scopolamine, place cells displayed relatively consistent firing locations. Although most of the place cells still showed a certain degree of place sensitivity after scopolamine injection, it was greatly reduced compared with baseline and the neural off-field firing became stronger. In order to quantify the level of "ensemble stability", we analysed the PVO between the odd trials and even trials data within every session. This revealed the degree of overlap of the ensemble activity in these two different conditions. In Figure 3.4, a clear diagonalization feature during the baseline assessment is demonstrated, showing a specific and consistent ensemble firing pattern in both even and odd trials. After scopolamine injection, this diagonalization feature was greatly reduced. The DI was significantly higher in saline treated mice (RL: 102.60% ± 5.61% and LR: 99.07% ± 4.46% relative to baseline) than scopolamine-treated mice (RL: 62.88% ± 4.64% and LR: 66.24% ± 4.74% relative to

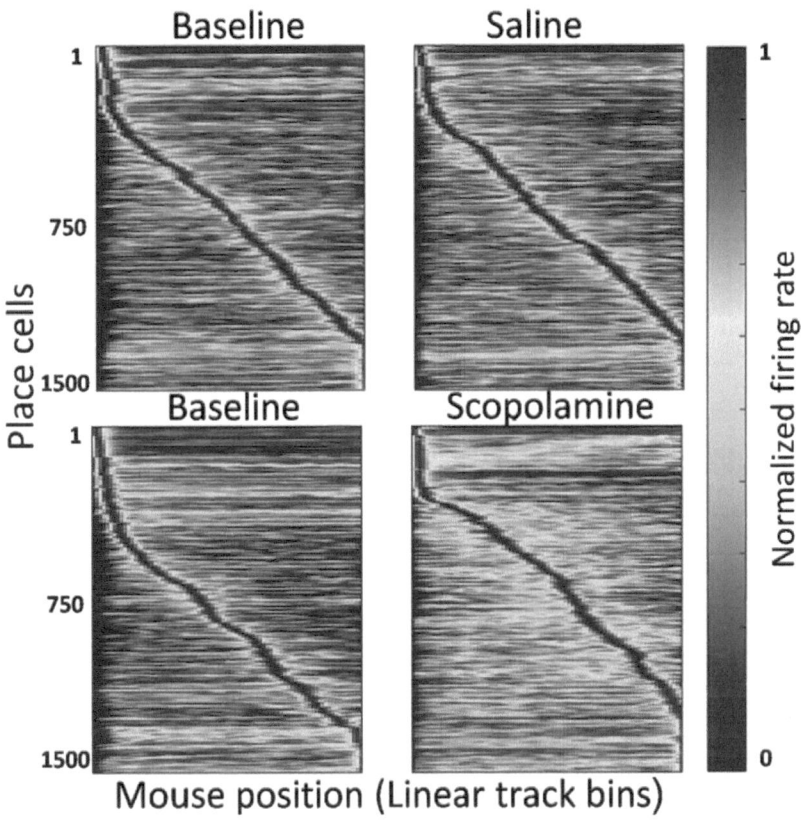

FIGURE 3.3 Normalized place field map when the animals run on the linear track before and after the administration of saline or scopolamine. The bin size was 2 cm. The location sensitivity of the place cells in saline group was very stable, while it decreased considerably after scopolamine administration.

baseline) (two-way repeated-measures ANOVA; main effect of injection, $F_{1,4} = 45.56$, $p = 0.003$; main effect of running direction, $F_{1,4} = 0.003$, $p = 0.956$; interaction, $F_{1,4} = 7.57$, $p = 0.051$; *post hoc* test saline versus scopolamine: RL, $p < 0.0001$; LR, $p < 0.0001$). The difference of the firing patterns was compared separately by using the autoencoder described above. After data dimensionality reduction and extraction of the most significant information from the input spike-train epochs, epochs in the scopolamine treated groups were clearly separated into two clusters, while the saline control group converged (Figure 3.5).

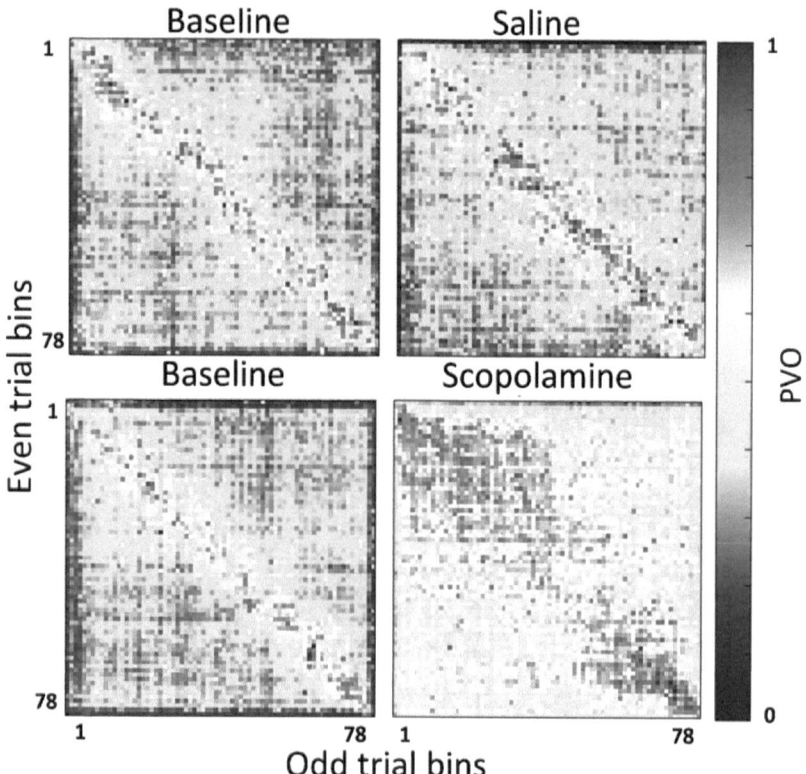

FIGURE 3.4 Normalized PVO of the place cells between odd trials and even trials before and after the administration of saline or scopolamine. The overlap along the diagonal in baseline indicated spiking at the same bins, and this feature disappeared after scopolamine injection.

3.3.4 Scopolamine Decreased Decoding Accuracy

We analysed the animal's neural decoding accuracy by using a Bayesian decoder to predict the animal's position (see "Methods"). Only the activity of place cells was used to train the decoder. The decoding error ratio in scopolamine group ($270.74\% \pm 26.48\%$ relative to baseline) was significantly higher than that in saline group ($98.79\% \pm 7.80\%$ relative to baseline; paired t-test, $p = 0.007$). Prior to the injection of scopolamine, the average estimation error was 1.45 ± 0.14 cm/frame, increasing to 4.02 ± 0.61 cm/frame following injection.

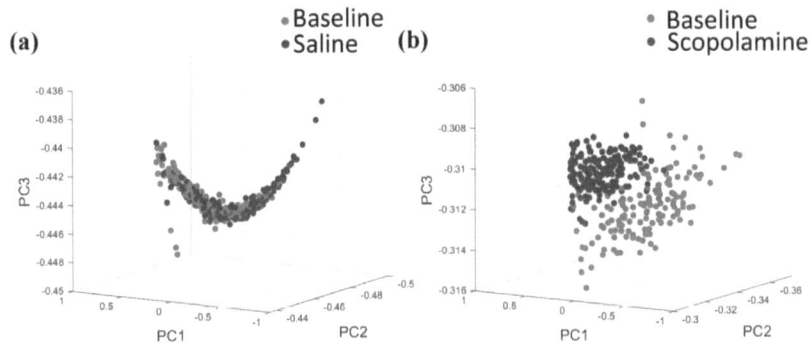

FIGURE 3.5 Examples of principal component space for the results of the spike-train epochs after applying unsupervised autoencoder. Each circle represents a 2.5 s data segment of recordings at baseline(red) or (a) saline / (b) scopolamine(blue).

3.3.5 Effect on Running Speed

The animal's average running speed was measured using all the data on both running directions (except the bins on each end of the linear track). The scopolamine treated mice had a slightly higher running velocity (114.32% ± 8.10% relative to baseline) compared with saline treated ones (100.01% ± 5.66% relative to baseline; paired t-test, $p = 0.02$).

3.3.6 A Pyramidal/Interneuron Conductance-Based Model of CA1 Place Cells Recapitulates Reduced Cell Firing and Place Cell Specificity through M-Current Modulation

Modelling scopolamine effects by enhancement of M-channel conductance impaired spatial firing patterns significantly with decreased neural firing rate, quantity of place cells, and SIC, leading to less-consistent spatial firing patterns (one-way ANOVA, $F_{4,20} = 3384$, $p < 0.001$; Figure 3.6a). The percentage of place cells in the control group was 82.47% ± 1.43%, while it decreased to 66.92% ± 1.26% in ×5 group ($p < 0.0001$); 55.84% ± 0.83% in ×10 group ($p < 0.0001$); 42.61% ± 0.93% in ×15 group ($p < 0.0001$) and 30.15% ± 0.96% in ×20 group ($p < 0.0001$). Similarly, the SIC also showed a decreasing trend with respect to the enhanced M-channel conductance (one-way ANOVA, $F_{4,20}$ = 337.6, p < 0.0001). SIC was 2.28 ± 0.02 bits in both control and × 5 (p > 0.99) groups, but it decreased to 1.98 ± 0.01 bits in × 10 group (p < 0.0001); 1.77 ± 0.02 bits in × 15 group (p < 0.0001) and 1.69 ± 0.02 bits in × 20 group (p < 0.0001). In Figure 3.6b, a clear diagonalization feature was observed in the control group, while this feature was greatly reduced in the other groups (one-way ANOVA, $F_{4,20}$ = 132.8, p < 0.0001). The DI was $0.66 \times 10^{-1} \pm 0.54 \times 10^{-3}$ in the control group, significantly reduced to $0.55 \times 10^{-1} \pm 0.43 \times 10^{-3}$ in ×5 group (83.09% ± 1.00% of the control level, p < 0.0001), $0.53 \times 10^{-1} \pm 0.13 \times 10^{-3}$ in ×10 group (80.24% ± 1.59% of the control level, p < 0.0001), $0.46 \times 10^{-1} \pm 0.89 \times 10^{-3}$ in × 15 group (69.62% ± 1.27% of the control level, p < 0.0001), and $0.42 \times 10^{-1} \pm 0.49 \times 10^{-3}$ in × 20 group (64.47% ± 0.83% of the control level, p < 0.0001).

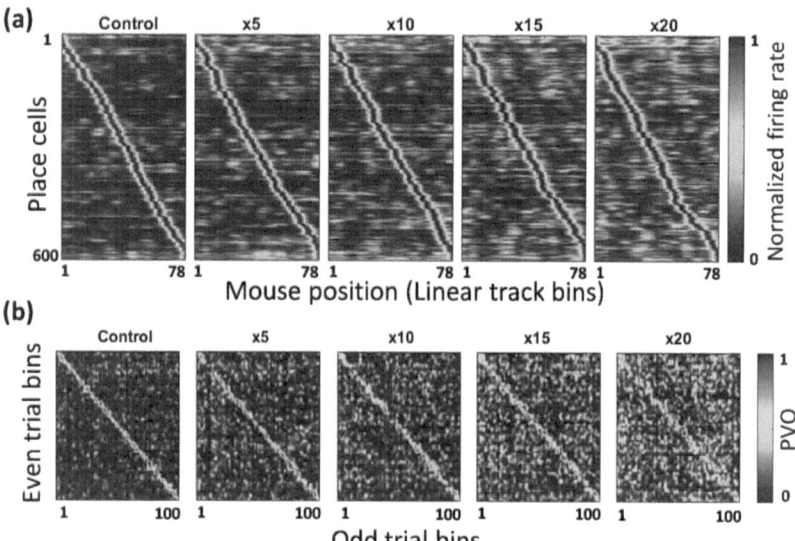

FIGURE 3.6 M-channel conductances enhancement in a simulated hippocampal network model replicated experiment results. (a) The decreased neural spatial sensitivity was observed in the place field maps with enhancements of M-channel conductances of pyramidal cells by five (×5), ten (×10), fifteen (×15), and twenty (×20) times separately. The M-channel conductances of SST+ and PV+ interneurons were enhanced to a level reproducing the neural firing reduction observed experimentally. (b) The "diagonalization feature" of ×5, ×10, ×15, and ×20 groups significantly reduced compared with the control group.

3.4 DISCUSSION

We have studied the effect of systemic administration of scopolamine on the activity of hippocampal place cells in mice running in a linear track. The main effects observed were a notable reduction in the activity of neurons throughout the CA1 region, as well as greatly impaired spatial resolution and associated metrics of place cells.

3.4.1 The Effects of ACh on Memory

Previous studies have hypothesized that Ach facilitates memory encoding through enhancement of the signal-to-noise ratio, improving theta rhythm modulation, increasing persistent spiking or potentiating afferent input[10,24]. Pharmacological blockade of Ach transmission by mAChR antagonists atropine or scopolamine impairs the acquisition of new memories. In this study, we did not attempt to study encoding events, but rather the effects of mAChR blockade on neural activity as well as ensembles that have previously encoded spatial information in CA1, using miniscope neuronal calcium signals. Scopolamine had quite striking effects on cellular and ensemble behaviour and disrupted place cell specificity. The animals have been trained to traverse the linear track for multiple times before the experiment and given 30 min to become familiar with the environment again before the recording. The spatial memory should have already been fully encoded by previous training as the results of the neural firing rate, SIC, place cell quantity, decoding accuracy, and the PVO in the saline control group are quite stable before and after the injection. We did not formally behaviourally test recall but given the disruption to the encoded spatial information demonstrated, we expect recall would be likely to be impaired to some extent. While the animals were still capable of recalling the spatial position of the food reward, the retention of this basic ability does not exclude the possibility of some degradation of recall. This could be validated with the use of more accurate methods such as the Morris or Barnes maze. Huang et al. documented impairment of memory recall in scopolamine treated mice in the Morris maze[12], consistent with this hypothesis.

3.4.2 The Effects of mAChR Blockade on Neural Cellular Activity and Network Dynamics

The reduction in both the quantity of detected cells, average neuron firing rate, and cell properties implied impairment of neuronal activity at the network level likely related to the pharmacological effects of muscarinic blockade. One mechanism by which the reduction in cellular activity could be explained is by the effect of muscarinic receptors on potassium channels, especially the M-current. The M-current is carried by Kv7 subtype channels inhibited by muscarinic receptors that are expressed at high density at the axon hillock – a specialized area of neurons modulating cell firing[25]. ACh normally inhibits this current and increases neuronal firing. Hence, scopolamine would be expected to reduce overall neural activity. Previous studies have identified the role of ACh in regulating neural excitability by modulating the M-current[26]. There are other mechanisms by which muscarinic receptors could decrease neuronal activity by enhancing potassium conductance, such as the calcium-activated non-selective cationic current, which is likely to modulate neural persistent firing but less strongly than M-current[27]. The place field maps became much less precise after the administration of scopolamine, with increased neuronal activity outside the place field. This is in consistent with the results of Brazhnik et al. using electrophysiologically identified CA1 place cells in rats[28]. Newman et al. found that scopolamine significantly reduced phase precession of spiking relative to the local field theta and the spatial information of the cells, but the spatial tuning of the cells that had preserved place fields remained stable[29]. In the current experiments, we found that there was greatly increased variability of neuronal activity in the observed neurons, both in place fields and surrounding areas which was quantified by calculating the PVO between odd and even trials. This variability was manifested as greatly reduced "diagonalization" after blocking mAChRs, indicating weakened temporal stability. Additionally, these changes were further explored with the autoencoder algorithm. We found that the encoder effectively separated the spike-train epochs before and after mAChRs blockade, implying the effects of scopolamine on the neural network may be reflected in a higher dimensional metric space. To further explore the impairment of ensemble spatial encoding, we used a naive Bayesian classifier to estimate the animal's position in the linear track. After the administration of scopolamine, the position estimation deteriorated with a great increase in error rate, consistent with the PVO changes.

3.4.3 The Effects of ACh on Animal Behaviour

We used time of onset and drug doses similar to previous studies[9]. We found the average running speed increased slightly after injecting scopolamine, differing from previous findings[29]. This may be due to a higher range dose used, complex pharmacological effects of muscarinic blockade, or possibly a difference between mice and rats. An increase in locomotion speed was observed in Bushnell (1987) and Thomsen (2014) with dose-dependent enhancement in mobility, with about 25% and 50% increase with 1 mg/kg respectively[30-31].

3.4.4 Hippocampal CA1 Network Model

Place cells were simulated based on previously described hippocampal network models[23]. In these models, the synaptic input to the network originates from EC, CA3, and medial septum. As low ACh level is believed to shift network dynamics toward memory retrieval and suppress the memory encoding process[32], we deleted the EC-CA1 connectivity in our model to simulate the retrieval of spatial information. The blockade of mAChR normally enhances the M-current at the cellular level, but exceptions have also been reported[33]. In the simulation study, we hypothesized that a major effect of scopolamine would be the enhancement of the M-current. Inhibition of the M-current has been identified to modulate neuronal firing rates in pyramidal cells[34]. In hippocampal interneurons, ACh affects firing properties. Both PV+ and SST+ interneurons produce a range of responses to muscarinic agonists[35-36]. Immunohistochemistry results show the expression of M-channel on PV+ interneurons[37], and mAChRs are observed on PV+ interneurons by Cea-del et al. (2010)[38]. The model reproduced the reduced place cell number and information content, sparser place field map and impaired stability of spatial representation revealed by hippocampal neural ensemble (PVO map). Differences occurred between the model-derived place field map and the one that empirically observed, possibly due to the simplifications inherent in the model.

3.4.5 Miniscope Methodology

Electrophysiological studies commonly use single electrode or electrode arrays to study the activities of brain networks with the help of associated spike sorting algorithms. These algorithms are usually based on the properties of the neural firing rate, spike duration, and the autocorrelation function[39] in order to distinguish the spikes generated by different neurons. Hence, spatial resolution is not always guaranteed. Another issue with this technique is the limited quantity of cells detected and potential variability over time. In contrast, the miniscope provides a larger area of view, generally displaying the stable activity of hundreds of CA1 neurons that is particularly suitable for the study of network activity. Nonetheless, the slow dynamics of the calcium activity, the temporal resolution of calcium imaging cannot match that of electrophysiological recording. Pharmacological manipulation by muscarinic antagonists such as scopolamine might possibly directly affect the calcium conductance of the neurons. A relevant question is whether the effects seen with calcium imaging may just result from the impact of muscarinic antagonism on calcium transients, rather than neuronal activity as such. Interestingly, muscarinic agonists have been found to inhibit calcium currents in hippocampal pyramidal cells[40], and the calcium signal in response to a single action potential is reduced during muscarinic stimulation[41], implying muscarinic blockade could potentially enhance calcium signal in some circumstances. Nonetheless, several studies have shown that muscarinic agonists enhance intracellular calcium transients, especially in dendrites and with multiple action potentials[41], potentially causing the apparent reduction in neuronal activity we report. While it is impossible to absolutely exclude such direct effects, we think it is likely the cellular activity recorded by the miniscope does substantially reflect true neuronal and network activity affected by scopolamine, as similar effects were seen with electrophysiological recordings by Brazhnik et al. (2004)[28]. The virus injected into hippocampal CA1 is non-specifically expressed in both pyramidal cells and interneurons; both play essential roles in the animal's spatial navigation via different mechanisms of action[42]. It would be of great interest to observe interneuron dynamics separately in response to scopolamine. However, the current miniscope recording system cannot technically distinguish them. A possible way to overcome this issue is to inject another specific-targeting virus to mark the interneurons with a different colour of fluorophore and replace the stimulation monocolour light source to multicolour[43]. Both enhancement of the recording system and experimental design deserve further study.

In summary, our results demonstrate that mAChRs have an important role in regulating spatial memory and scopolamine has a strong disruptive effect on the decoding of neural correlates of spatial memory, most likely related to the blockade of mAChR. This is plausibly modulated through increasing M-current amplitude with reduction of neuronal excitability and impairment of network function. Taken as a whole, our results are in accordance with the Hasselmo's proposal (2006) of a primary effect of Ach being to enhance the signal-to-noise ratio in neural networks. It will be particularly interesting in future studies to directly observe interneuron dynamics, as well as directly assess spatial information decoding.

REFERENCES

1. Dolan, R.J., and Fletcher, P.C. (1997). Dissociating prefrontal and hippocampal function in episodic memory encoding. *Nature* 388, 582–585. doi: 10.1038/41561.

2. Carr, M.F., Jadhav, S.P., and Frank, L.M. (2011). Hippocampal replay in the awake state: a potential substrate for memory consolidation and retrieval. *Nat Neurosci* 14, 147–153. doi: 10.1038/nn.2732.

3. O'keefe, J., and Dostrovsky, J. (1971). The hippocampus as a spatial map. Preliminary evidence from unit activity in the freely-moving rat. *Brain Res* 34, 171–175. doi: 10.1016/0006-8993(71)90358-1.

4. O'keefe, J., and Nadel, L. (1978). *The Hippocampus as a Cognitive Map.* Oxford: Clarendon Press.

5. Pfeiffer, B.E., and Foster, D.J. (2013). Hippocampal place-cell sequences depict future paths to remembered goals. *Nature* 497, 74–79. doi: 10.1038/nature12112.

6. Zhu, H., Yan, H., Tang, N., Li, X., Pang, P., Li, H., et al. (2017). Impairments of spatial memory in an Alzheimer's disease model via degeneration of hippocampal cholinergic synapses. *Nat Commun* 8, 1676. doi: 10.1038/s41467-017-01943-0.

7. Hayward, P. (2004). Acetylcholine and memory formation. *The Lancet Neurology* 3, 201.

8. Mangialasche, F., Solomon, A., Winblad, B., Mecocci, P., and Kivipelto, M. (2010). Alzheimer's disease: clinical trials and drug development. *Lancet Neurol* 9, 702–716. doi: 10.1016/S1474-4422(10)70119-8.

9. Klinkenberg, I., and Blokland, A. (2010). The validity of scopolamine as a pharmacological model for cognitive impairment: a review of animal behavioral studies. *Neurosci Biobehav Rev* 34, 1307–1350. doi: 10.1016/j.neubiorev.2010.04.001.

10. Hasselmo, M.E. (2006). The role of acetylcholine in learning and memory. *Curr Opin Neurobiol* 16, 710–715. doi: 10.1016/j.conb.2006.09.002.

11. Dong Hyun, K., and Ryu, J.H. (2008). Differential effects of scopolamine on memory processes in the object recognition test and the Morris water maze test in mice. *Biomol Therap* 16, 173–178.

12. Huang, Z.B., Wang, H., Rao, X.R., Zhong, G.F., Hu, W.H., and Sheng, G.Q. (2011). Different effects of scopolamine on the retrieval of spatial memory and fear memory. *Behav Brain Res* 221, 604–609. doi: 10.1016/j.bbr.2010.05.032.

13. Ghosh, K.K., Burns, L.D., Cocker, E.D., Nimmerjahn, A., Ziv, Y., Gamal, A.E., et al. (2011). Miniaturized integration of a fluorescence microscope. *Nat Methods* 8, 871–878. doi: 10.1038/nmeth.1694.

14. Pnevmatikakis, E.A., and Giovannucci, A. (2017). NoRMCorre: An online algorithm for piecewise rigid motion correction of calcium imaging data. *J Neurosci Methods* 291, 83–94. doi: 10.1016/j.jneumeth.2017.07.031.

15. Zhou, P., Resendez, S.L., Rodriguez-Romaguera, J., Jimenez, J.C., Neufeld, S.Q., Giovannucci, A., et al. (2018). Efficient and accurate extraction of in vivo calcium signals from microendoscopic video data. *Elife* 7. doi: 10.7554/eLife.28728.

16. Friedrich, J., Zhou, P., and Paninski, L. (2017). Fast online deconvolution of calcium imaging data. *PLoS Comput Biol* 13, e1005423. doi: 10.1371/journal.pcbi.1005423.

17. Rubin, A., Geva, N., Sheintuch, L., and Ziv, Y. (2015). Hippocampal ensemble dynamics timestamp events in long-term memory. *Elife* 4. doi: 10.7554/eLife.12247.
18. Ravassard, P., Kees, A., Willers, B., Ho, D., Aharoni, D.A., Cushman, J., et al. (2013). Multisensory control of hippocampal spatiotemporal selectivity. *Science* 340, 1342–1346. doi: 10.1126/science.1232655.
19. Kramer, and Kramer, M. (1991). Nonlinear principal component analysis using autoassociative neural networks. *AIChE Journal* 37, 233–243.
20. Abadi, M. (2016). *TensorFlow: A System for Large-Scale Machine Learning.* In *12th USENIX Symposium on Operating Systems Design and Implementation (OSDI 16)* (pp. 265–283).
21. Davidson, T.J., Kloosterman, F., and Wilson, M.A. (2009). Hippocampal replay of extended experience. *Neuron* 63, 497–507. doi: 10.1016/j.neuron.2009.07.027.
22. Turi, G.F., Li, W.K., Chavlis, S., Pandi, I., O'hare, J., Priestley, J.B., et al. (2019). Vasoactive intestinal polypeptide-expressing interneurons in the hippocampus support goal-oriented spatial learning. *Neuron* 101, 1150–1165 e1158. doi: 10.1016/j.neuron.2019.01.009.
23. Shuman, T., Aharoni, D., Cai, D.J., Lee, C.R., Chavlis, S., Page-Harley, L., et al. (2020). Breakdown of spatial coding and interneuron synchronization in epileptic mice. *Nat Neurosci* 23, 229–238. doi: 10.1038/s41593-019-0559-0.
24. Dannenberg, H., Young, K., and Hasselmo, M. (2017). Modulation of hippocampal circuits by muscarinic and nicotinic receptors. *Front Neural Circuits* 11, 102. doi: 10.3389/fncir.2017.00102.
25. Shah, M.M., Migliore, M., Valencia, I., Cooper, E.C., and Brown, D.A. (2008). Functional significance of axonal Kv7 channels in hippocampal pyramidal neurons. *Proc Natl Acad Sci U S A* 105, 7869–7874. doi: 10.1073/pnas.0802805105.
26. Cole, A.E., and Nicoll, R.A. (1983). Acetylcholine mediates a slow synaptic potential in hippocampal pyramidal cells. *Science* 221, 1299–1301. doi: 10.1126/science.6612345.
27. Knauer, B., Jochems, A., Valero-Aracama, M.J., and Yoshida, M. (2013). Long-lasting intrinsic persistent firing in rat CA1 pyramidal cells: a possible mechanism for active maintenance of memory. *Hippocampus* 23, 820–831. doi: 10.1002/hipo.22136.
28. Brazhnik, E., Borgnis, R., Muller, R.U., and Fox, S.E. (2004). The effects on place cells of local scopolamine dialysis are mimicked by a mixture of two specific muscarinic antagonists. *J Neurosci* 24, 9313–9323. doi: 10.1523/JNEUROSCI.1618-04.2004.
29. Newman, E.L., Venditto, S.J.C., Climer, J.R., Petter, E.A., Gillet, S.N., and Levy, S. (2017). Precise spike timing dynamics of hippocampal place cell activity sensitive to cholinergic disruption. *Hippocampus* 27, 1069–1082. doi: 10.1002/hipo.22753.
30. Bushnell, P.J. (1987). Effects of scopolamine on locomotor activity and metabolic rate in mice. *Pharmacol Biochem Behav* 26, 195–198. doi: 10.1016/0091-3057(87)90555-7.

31. Thomsen, M. (2014). Locomotor activating effects of cocaine and scopolamine combinations in rats: isobolographic analysis. *Behav Pharmacol* 25, 259–266. doi: 10.1097/FBP.0000000000000043.
32. Hasselmo, M.E. (1999). Neuromodulation: acetylcholine and memory consolidation. *Trends Cogn Sci* 3, 351–359. doi: 10.1016/s1364-6613(99)01365-0.
33. Carver, C.M., and Shapiro, M.S. (2019). Gq-coupled muscarinic receptor enhancement of KCNQ2/3 channels and activation of TRPC channels in multimodal control of excitability in dentate gyrus granule cells. *J Neurosci* 39, 1566–1587. doi: 10.1523/JNEUROSCI.1781-18.2018.
34. Honigsperger, C., Marosi, M., Murphy, R., and Storm, J.F. (2015). Dorsoventral differences in Kv7/M-current and its impact on resonance, temporal summation and excitability in rat hippocampal pyramidal cells. *J Physiol* 593, 1551–1580. doi: 10.1113/jphysiol.2014.280826.
35. Xiang, Z., Huguenard, J.R., and Prince, D.A. (1998). Cholinergic switching within neocortical inhibitory networks. *Science* 281, 985–988. doi: 10.1126/science.281.5379.985.
36. Mcquiston, A.R., and Madison, D.V. (1999). Muscarinic receptor activity induces an afterdepolarization in a subpopulation of hippocampal CA1 interneurons. *J Neurosci* 19, 5703–5710.
37. Cooper, E.C., Harrington, E., Jan, Y.N., and Jan, L.Y. (2001). M channel KCNQ2 subunits are localized to key sites for control of neuronal network oscillations and synchronization in mouse brain. *J Neurosci* 21, 9529–9540.
38. Cea-Del Rio, C.A., Lawrence, J.J., Tricoire, L., Erdelyi, F., Szabo, G., and Mcbain, C.J. (2010). M3 muscarinic acetylcholine receptor expression confers differential cholinergic modulation to neurochemically distinct hippocampal basket cell subtypes. *J Neurosci* 30, 6011–6024. doi: 10.1523/JNEUROSCI.5040-09.2010.
39. Rossant, C., Kadir, S.N., Goodman, D.F.M., Schulman, J., Hunter, M.L.D., Saleem, A.B., et al. (2016). Spike sorting for large, dense electrode arrays. *Nat Neurosci* 19, 634–641. doi: 10.1038/nn.4268.
40. Gahwiler, B.H., and Brown, D.A. (1987). Muscarine affects calcium-currents in rat hippocampal pyramidal cells in vitro. *Neurosci Lett* 76, 301–306. doi: 10.1016/0304-3940(87)90419-8.
41. Power, J.M., and Sah, P. (2002). Nuclear calcium signaling evoked by cholinergic stimulation in hippocampal CA1 pyramidal neurons. *J Neurosci* 22, 3454–3462. doi: 20026335.
42. Gloveli, T. (2010). Hippocampal spatial navigation: interneurons take responsibility. *J Physiol* 588, 4609–4610. doi: 10.1113/jphysiol.2010.200212.
43. Aharoni, D., Khakh, B.S., Silva, A.J., and Golshani, P. (2019). All the light that we can see: a new era in miniaturized microscopy. *Nat Methods* 16, 11–13. doi: 10.1038/s41592-018-0266-x.

Hippocampal Cognitive and Relational Map Paradigms Explored by Multisensory Encoding Recording with Widefield Calcium Imaging

4

4.1 INTRODUCTION

In addition to being used in the medical research area, another important use of OBCIs is to study the animal's sensitivity to the external world. The conventional way of using BCI is to apply it in a specific region to decode a particular sensory modality. For example, one might use visual cortex neuronal activity to decode visual information and auditory cortex signals to decode auditory information. One challenge is to find a region that not only integrates information from multiple sensory modalities but also ensures that this information

DOI:10.1201/9781003470397-4

is decodable. In this chapter, we will describe a study that examines the hippocampal multisensory encoding of spatial, visual, and auditory information. Additionally, we will introduce more advanced analysis methods to analyse the signals recorded by OBCI, such as manifold analysis, criticality analysis, and neural network topology analysis.

The hippocampus plays a crucial role in memory, navigation, and cognition[1-3]. Previous studies have developed several theories of rodent hippocampal function, which can be divided into two distinct categories depending on the role assigned to spatial information processing. The influential *cognitive map* theory views spatial information as the primary organizational principle, while signals of other sensory modalities are integrated in a spatial framework as features for processing[4,5]. In contrast, the *relational* theory proposes that the hippocampus supports a broader function with different types of information processed independently, and spatial representation is simply an example of a general mechanism for encoding information[1,6,7]. It has been shown that non-spatial factors can modulate the activity of hippocampal neurons, such as olfactory, gustatory, tactile, auditory, and visual. However, these non-spatial responses have been found to be spatial context- or task-dependent in almost all previous studies[5]. Consequently, such responses may actually represent the location or context in which the feature is present rather than the feature per se, and in turn challenge the relational theory.

However, no experiments have been designed to exclude the interference of spatial information when studying non-spatial responses. It has become apparent that responses to salient stimuli are more adequately and stably represented by neural population activity compared with intrinsically variable single neuron responses[8]. Additionally, it has been reported that multidimensional population responses can be more effectively parsed using population-level analyses such as graph topological analyses or manifold analyses[9,10], which have not previously been employed for analysis non-spatial information in CA1. Interestingly, it has recently been reported that rhythmic light stimulation can enhance memory function in a rodent model of Alzheimer's disease[11], suggesting more complex roles for light stimulation beyond simple sensory detection and emphasizing the need for a more comprehensive understanding of hippocampal responses to non-spatial stimuli.

Since it is impossible to eliminate all spatial inputs to the animal, we placed the animal in a small recording chamber during the experiment (Figure 4.1). This arrangement was designed to minimize spatial influences, allowing the animal to be passively exposed to either visual stimuli, auditory stimuli at different frequencies, or a combination of both. Notably, stimuli were introduced randomly throughout the experiment, a factor that significantly reduced the interference of spatial inputs, because if neuronal responses do indeed represent feature-in-place responses, consistent stimuli evoked neuronal firing patterns should not

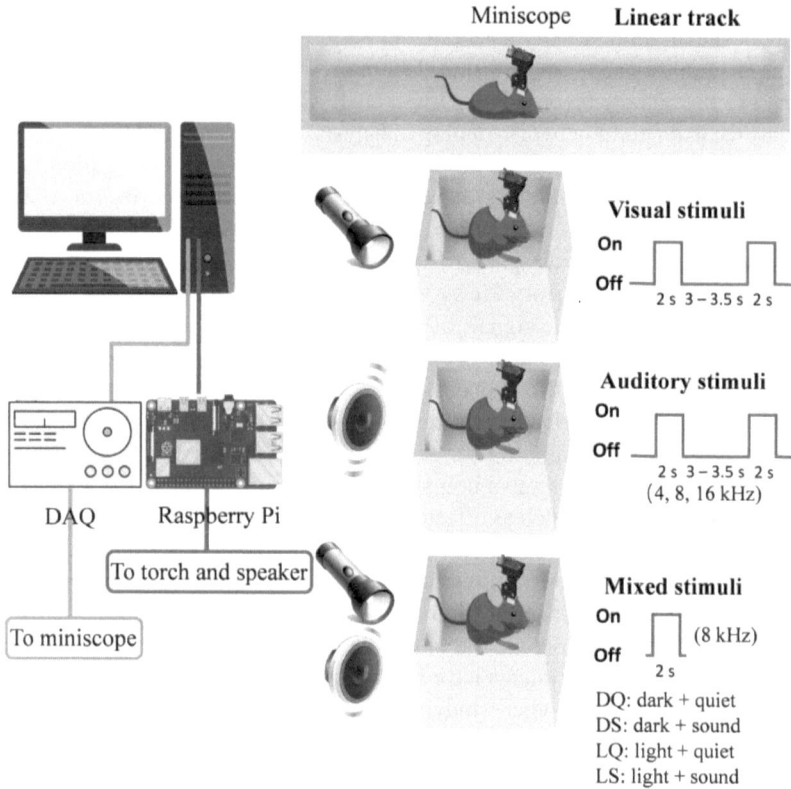

FIGURE 4.1 Hippocampal calcium activity and experimental design. The animal's brain activity was studied in four separate experiments: (1) mice traversed a linear track; mice were exposed to either (2) light stimuli or (3) sound stimuli at different frequencies or (4) light-sound mixed stimuli. The torch and speaker were controlled using a Raspberry Pi board that was synchronized with the miniscope recording system.

be detectable. Additionally, we observed the population activity when animals traversed a linear track. The hippocampal calcium activity was recorded simultaneously using a miniaturized fluorescence microscope[12] ("miniscope"). We first studied the properties of individual neuronal activity, and then topology connectivity and population activity using graph theory and manifold analysis, respectively. Additionally, we tested for critical dynamical behaviour of neuronal populations over time to determine neural dynamic variation in response to repeated stimuli. We demonstrate that the hippocampus processes non-spatial information irrespective of spatial inputs. The network topology shows very dense connections for spatial information but quite sparse connections for non-spatial information, and neuronal populations displayed different coding mechanisms depending on sensory inputs. The neuronal system moves away from the critical state over time in visual and auditory stimuli experiments but remained stable in mixed stimuli experiments. Together, these results enhance our understanding of hippocampal neuronal population activity in response to various modalities of information and provide support for the relational theory.

4.2 RESULTS

Hippocampal neural activity was studied in four separate experimental scenarios: light stimuli, sound stimuli, linear track, and light-sound mixed stimuli. We analysed both the neuronal firing patterns of each individual neuron and the population activity. The initial part of our analysis primarily focused on assessing the informational characteristics of each neuron's firing pattern. Information content was used to quantify the amount of information encoded within neuronal firing patterns, aiding in the evaluation of neural coding complexity. Additionally, we measured the cosine similarity index to evaluate the stability of these firing patterns across trials. Furthermore, mutual information was measured to assess the statistical dependencies between different stimuli and to evaluate the flow of information. Finally, we utilized a machine learning-based autoencoder method to uncover hidden deep features within the neuronal firing patterns of each neuron. The second part of the analysis focused on population activity. Graph theory modelled complex connectivity and interactions between neuronal populations in the network, while manifold analysis explored high-dimensional neuronal data structures to identify the underlying population firing patterns. Criticality analysis provided insights into neural network stability and adaptability. The comparison of the results in different experiments will be elaborated later in the discussion section.

4.2.1 Neural Representation During Light Exposure

To explore the light sensitivity of hippocampal neuronal ensembles, we studied hippocampal calcium activity when mice ($n = 5$) were exposed to repeated light stimuli as described in the methods. Many neurons displayed clear stimuli evoked responses, especially during dark-to-light ($13.65\% \pm 1.92\%$) and light-to-dark transients ($23.08\% \pm 5.47\%$; Figure 4.2A). We then investigated the resulting neural activity in four temporal windows: 15 frames before and after time point 0 s (DL: dark-to-light transient), 1 s (L: light), 2 s

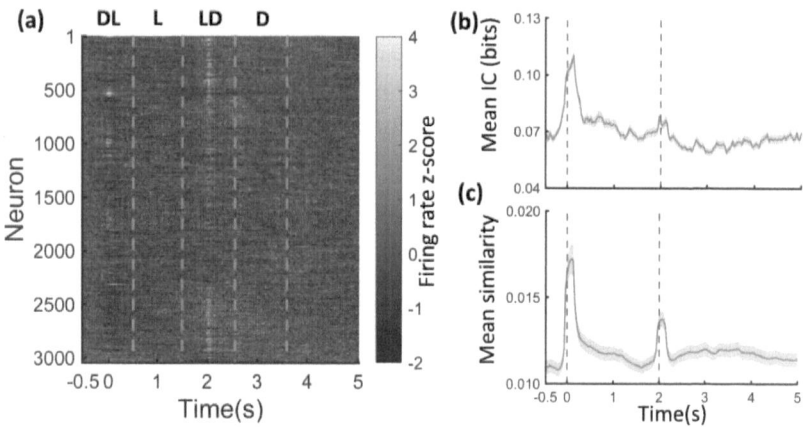

FIGURE 4.2 Both individual signal neurons and neuronal populations exhibited time-varying responses to visual stimuli. The flashlight was switched on at time $t = 0$ s and switched off at time $t = 2$ s. The activity was divided into four temporal windows for analysis: dark-to-light transient (DL), light (L), light-to-dark transient (LD), and dark (D). (A) The average neuronal temporal firing rate map over trials. (B) Neurons within the DL transient carried higher information content than others. (C) The temporal neuronal firing patterns within DL and LD transients showed higher stability across trials when measured using the cosine similarity index.

(LD: light-to-dark transient), and 3 s (D: dark). To assess the neuronal coding precision, we measured the information content of each neuron in four temporal windows. Neurons in the DL window had much higher information content, suggesting a more complex and precise coding process (Figure 4.2B). We next measured the cosine similarity index of each neuron to evaluate the temporal stability of neural firing patterns. Patterns in DL and LD windows demonstrated very high stability, especially around the activation and cessation of the stimulus. In contrast, patterns in L and D windows displayed comparable low stability (Figure 4.2C). To detect non-obvious activity patterns, we used an autoencoder to reduce the data dimensionality and extracted the most significant features. The autoencoder greatly improved discrimination, with clear distinctions detected even in L and D windows, where the above metrics did not reveal clear differences (Figure 4.3). We next measured the weighted normalized mutual information (WNMI) of each neuron between every two windows (DL-L, DL-LD, DL-D, L-LD, L-D, LD-D), and found that there were more sensitive neurons between transient periods and fewer between non-transient periods (one-way ANOVA, $F_{5,20} = 23.22$, $p = 0.0003$, *post hoc* test: DL-LD vs LD-D, $p = 0.0007$, DL-LD vs L-D, $p = 0.0352$, DL-D vs LD-D, $p = 0.0058$).

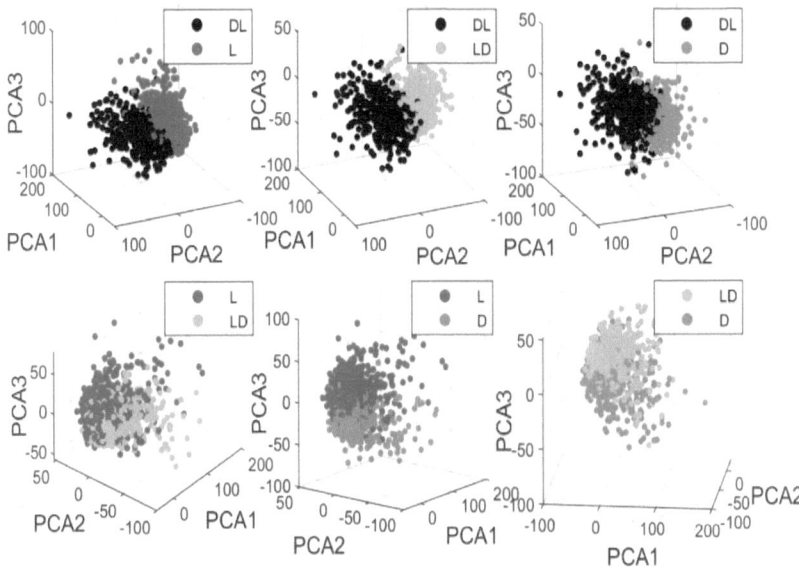

FIGURE 4.3 Examples of neuronal firing patterns analysed using an autoencoder. Each dot represents the firing pattern of a specific neuron. Since there were more than three extracted features, PCA was applied to the autoencoder results for better visualization.

Individual neurons exhibited clear stimuli-dependent firing patterns, but we sought to characterize the ensemble response as it is known that sets of neurons may encode salient properties in a more complex distributed pattern than evident from single neuronal responses. We therefore studied the topological connectivity and overall firing pattern of neuronal ensembles using graph theory and neural manifold analysis, respectively. We first measured correlation coefficients between neuron pairs and modelled neuronal population connectivity using graph theory. Network graphs of correlated neural activity demonstrated a clear stimuli-dependent topology. Compared with other windows, the network graph in the DL window had a denser structure, and there were more neurons showing stronger influence on the entire network (Figure 4.4A–B). The functional connectivity of the neuronal population varied, with neurons and cell groups in the DL window showing the highest clustering coefficient and more localized connectivity, respectively (Figure 4.4C). Significant connectivity differences were also detected between D and L windows (Figure 4.4D). We next studied overall population activity by seeking to identify neural manifolds. Cortical neuronal population activity displays redundancy in several systems, wherein multiple neuronal populations may exhibit analogous patterns of activity despite processing different types of

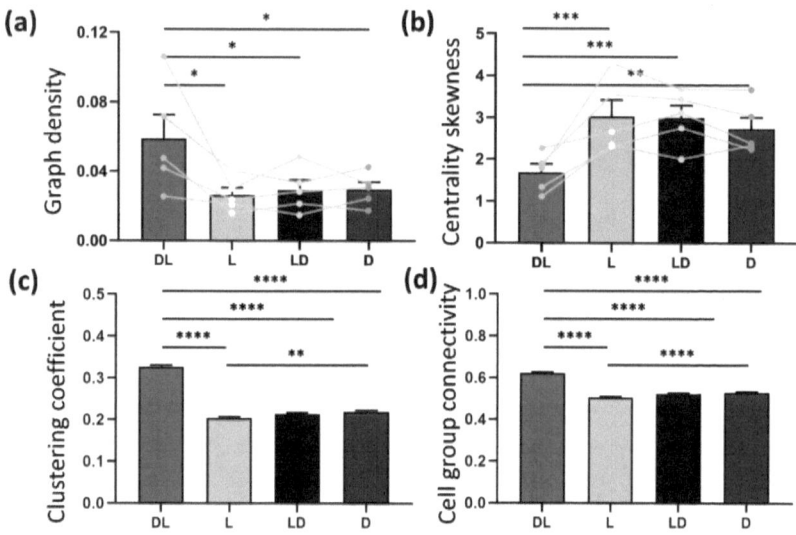

FIGURE 4.4 Neural network graph topology and neural manifolds in visual stimuli experiments. (A) Neurons in DL windows have higher graph density than others (one-way ANOVA, $F_{3,12}$ = 6.6, p = 0.0069; *post hoc* test: DL-L, p = 0.0133; DL-LD, p = 0.0257; DL-D, p = 0.0283). (B) There are more neurons with stronger influence on the network in DL windows than others (one-way ANOVA, $F_{3,12}$ = 14.2, p = 0.0003; post hoc test: DL-L, p = 0.0006; DL-LD, p = 0.0008; DL-D, p = 0.005). (C) Neurons in DL windows have stronger tendency to cluster together than others (one-way ANOVA, $F_{3,9132}$ = 403.2, p < 0.0001; *post hoc* test: DL-L/LD/D, p < 0.0001, L-D, p = 0.001). (D) Neurons in DL windows have more localized connectivity than others (one-way ANOVA, $F_{3,9132}$ = 177.4, p < 0.0001; *post hoc* test: DL-L/LD/D, p < 0.0001, L-D, p < 0.0001).

information or performing different tasks. Thus, we reduced the dimensionality of the population activity using a machine learning method (see Methods). We found that the majority of the variation in neuronal population activity (first three most significant features derived from the autoencoder) was represented by an intrinsic low-dimensional neural manifold that appeared confined to a closed-loop "C-shape" geometry (Figure 4.5A). Manifold amplitudes and angles provide two ways to quantify the similarity between neural modes and indicates when self-similarity breaks down. Throughout the onset and offset of the stimulus, the activity in neural manifolds revealed a substantial deviation from the origin of the coordinate. Additionally, the activity during light exposure displayed much larger angles than that during the non-exposure period, revealing changes in self-similarity of neural modes.

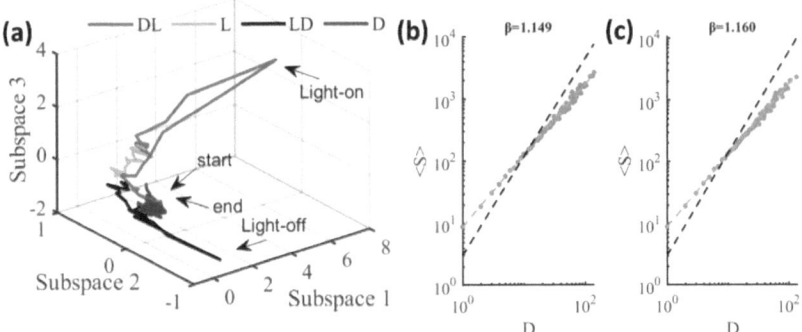

FIGURE 4.5 Neural manifolds representation and criticality analysis in visual stimuli experiments. (A) An example of the neural manifolds representation. (B–C) Criticality analysis of the neuronal population in visual stimuli experiment. Power–law distributions of the size (S), duration (D) in the (B) first half and (C) the second half of the experiment. The grey line represents the fitted β, and the red line represents the predicted β.

We next sought evidence for critical dynamical behaviour, and its temporal stability. We split the recording session in half and tested the population activity for markers of critical dynamics in each half. The system did display properties of criticality that varied over time. The sizes and durations of neuronal avalanches in both periods followed power–law distributions (Figure 4.5B–C). Several other measures of critical behaviour were also clearly demonstrated by the network behaviour including the deviation from criticality coefficient (DCC), branching ratio (BR), and shape collapse (SC) error, allowing more rigorous identification of critical behaviour. The network moved away from the critical state as time passed, as measured by the increased DCC error and reduced BR (DCC, Student's t-test, $p = 0.040$; BR, Student's t-test, $p = 0.032$). However, the SC errors did not appear to change significantly (Student's t-test, $p = 0.651$).

4.2.2 Neural Representation of Auditory Stimuli

We next studied neural activity when mice ($n = 5$) were exposed to sound stimuli at three different frequencies. The signal was sectioned into three types of epochs according to the stimuli frequency (4, 8, and 16 kHz) for analysis. We found that some neurons activated immediately after the onset of stimuli (4 kHz: 3.72% ± 2.22%, 8 kHz: 3.54% ± 1.04%, 16 kHz: 5.61% ± 2.54%), whereas most neurons started firing with varying degrees of delays

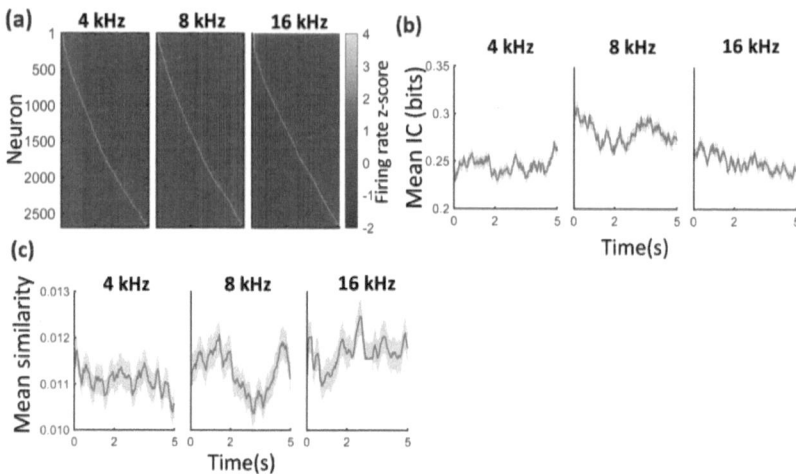

FIGURE 4.6 Distinct neuronal firing patterns were observed when the animal received auditory stimuli of different frequencies. The speaker was activated at time $t = 0$ s and deactivated at time $t = 2$ s. The activity was divided into three frequency epochs (4, 8, and 16 kHz) for analysis. (A) The average neuronal temporal firing rate map over trials in different frequency epochs. (B) Neurons in 8 kHz epochs carried higher information content than in other epochs. (C) Neuronal firing patterns in 16 kHz epochs showed the highest stability when measured using the cosine similarity index while patterns in 4 kHz epochs showed the lowest stability.

(Figure 4.6A). We then investigated coding properties of each neuron. Unlike visual stimuli experiments, neuronal information content remained comparatively stable over time without showing a surge with stimulus onset or offset. Compared to 4 and 16 kHz epochs, neurons in 8 kHz epochs displayed substantially higher information content (Figure 4.6B). Neuronal firing patterns remained temporally stable in 4 and 16 kHz epochs, with several neurons displaying strong stability immediately after stimulation. In contrast, the stability in 8 kHz epochs displayed a "U shape" pattern (Figure 4.6C). An autoencoder was applied to reduce the data dimensionality and identify hidden activity patterns. With the autoencoder, neuronal firing patterns in different frequency epochs could be distinguished into different clusters, suggesting that the sound frequency was encoded using quite different temporal coding processes (Figure 4.7). Intriguingly, there were fewer sensitive neurons when comparing the neural activity between 4 and 16 kHz epochs, indicating heterogeneity of hippocampal neurons to sound frequencies (Student's t-test; 4–8 kHz vs 4–16 kHz: $p < 0.0001$, 4–16 kHz vs 8–16 kHz: $p < 0.0001$).

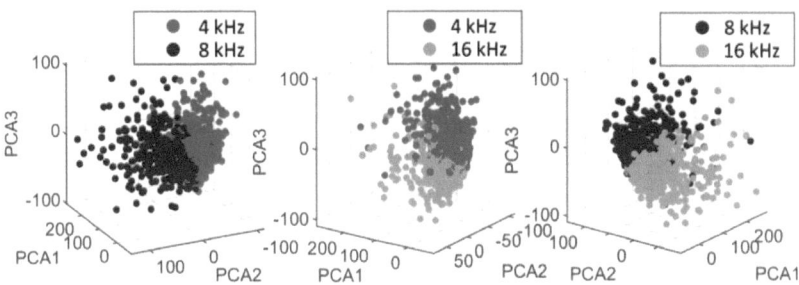

FIGURE 4.7 Examples of neuronal firing patterns analysed using an autoencoder. Each dot represents the firing pattern of a specific neuron. PCA was applied to results for better visualization.

We next characterized the neuronal population activity in response to sound and observed different frequency-dependent network topologies. Compared to 8 kHz epochs, neurons in 4 kHz epochs displayed a higher tendency to cluster (Figure 4.8C). Additionally, cell groups in 4 kHz epochs showed the strongest localized connectivity (Figure 4.8D). We also studied and compared the network density and the influence of each neuron to the network, but no significant differences were found with frequency (Figure 4.8A–B). In contrast to the visual stimuli findings, the neural manifold representing auditory frequency appeared to have a very "chaotic" geometry, but with each frequency confined to a separate subregion (Figure 4.9A). Population activity varied between scenarios, with manifold activity showing different angles. However, the manifold amplitude displayed quite similar patterns.

We then characterized criticality parameters of neuronal populations in the first half and the second half of the experiment. Again, the sizes and durations of neuronal avalanches in both periods were well approximated by truncated power–law distributions (Figure 4.9B–C). However, the DCC error significantly increased over time (DCC, Student's t-test, $p = 0.0027$; SC, Student's t-test, $p = 0.5500$; BR, Student's t-test, $p = 0.1316$), suggesting that the network gradually tuned away from the critical state over the period of observation.

4.2.3 Neural Representation During Linear Track Experiments

Non-spatial stimuli alter the characteristics of hippocampal CA1 neuronal ensembles. How do population activity and network topology change with spatial inputs? We next studied ensemble CA1 neural activity when well-trained

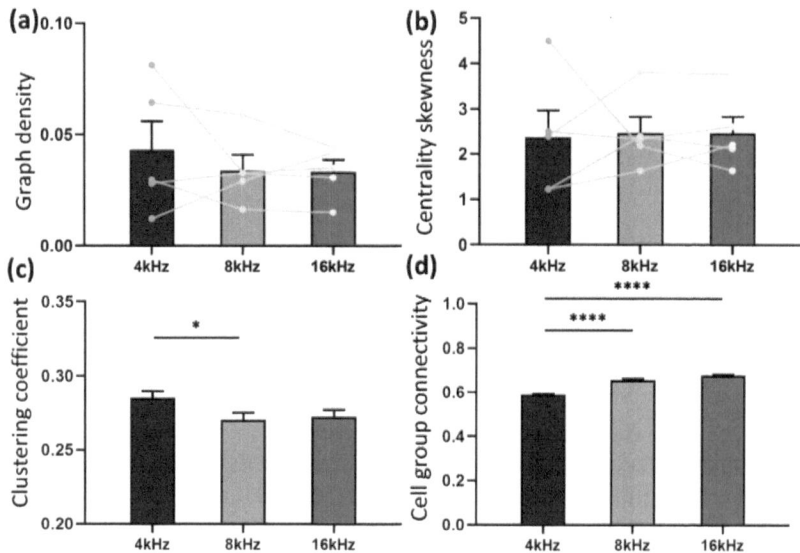

FIGURE 4.8 Neural network graph topology and neural manifolds in auditory stimuli experiments. (A) Density of the network in different epochs did not show significant differences (one-way ANOVA, $F_{2,8} = 0.6053$, $p = 0.5691$). (B) Network centrality distribution did not show significant differences (one-way ANOVA, $F_{2,8} = 0.0189$, $p = 0.9813$). (C) Neurons in 4 kHz windows have higher clustering coefficients than those in 8 kHz windows (one-way ANOVA, $F_{2,5376}=3.629$, $p = 0.0266$; *post hoc* test: 4–8 kHz, $p = 0.0397$). (D) Neurons in 4 kHz windows have more localized connectivity than others (one-way ANOVA, $F_{2,5376}=106.4$, $p < 0.0001$; *post hoc* test: 4–8 kHz/16 kHz, $p < 0.0001$).

mice ($n = 5$) traversed a linear track. The signal was divided into several 1-cm spatial bins, and we recorded neural responses in four long segments: spatial bins 1–40 (S1), 41–80 (S2), 81–120 (S3), and 121–160 (S4). Several neurons exhibited high information contents, particularly at both ends of the track (Figure 4.10A). Nonetheless, the stability of neuronal firing patterns was very low at both ends. On average, neurons in S4 had the lowest information content, but the highest stability (Figure 4.10B–C). However, after applying an autoencoder to extract significant features, activity in different segments could be separated into different clusters (Figure 4.11). Neurons showed different sensitivity in the four spatial segments. There were more sensitive neurons when comparing the activity between S1 and S4, and less-sensitive neurons between S2 and S3 (one-way ANOVA, $F_{5,20} = 7.657$, $p < 0.0004$; *post hoc*

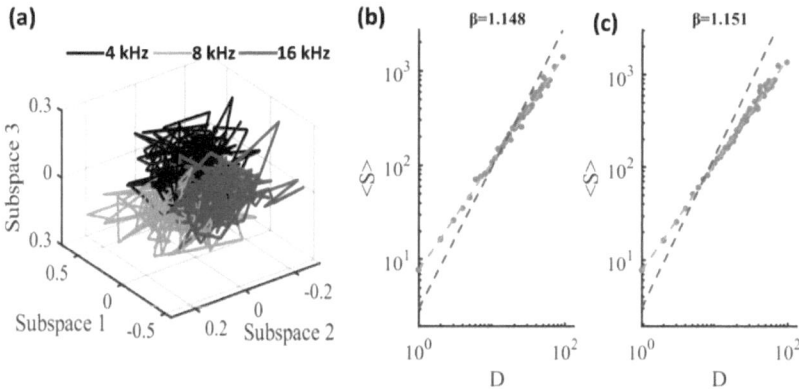

FIGURE 4.9 Neural manifolds representation and criticality analysis in auditory stimuli experiments. (A) An example of the neural manifolds representation. (B–C) Criticality analysis of the neuronal population in auditory stimuli experiment. Power–law distributions of the size (S), duration (D) in the (B) first half and (C) the second half of the experiment. The grey line represents the fitted β, and the red line represents the predicted β.

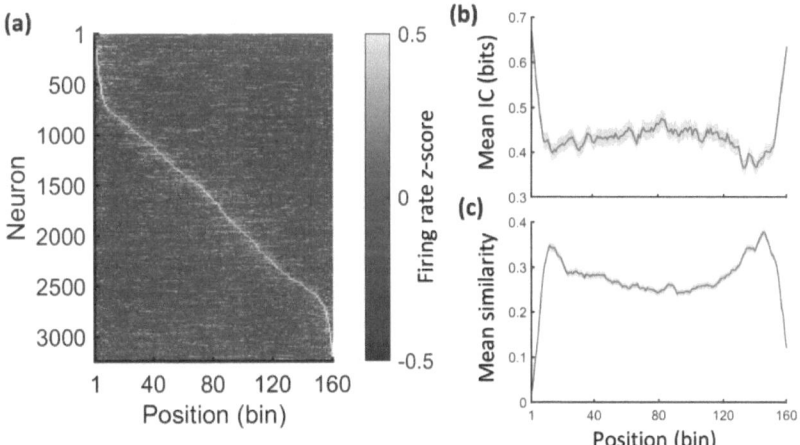

FIGURE 4.10 Hippocampal neuronal populations displayed distinct firing patterns in four long spatial segments: spatial bins 1–40 (S1), 41–80 (S2), 81–120 (S3), and 121–160 (S4). (A) Hippocampal place field map. (B) Neurons in S1 and S4 carried higher information content than that in S2 and S3. (C) Neuronal firing patterns in S1 and S4 showed high stability across trials when measured using the cosine similarity index.

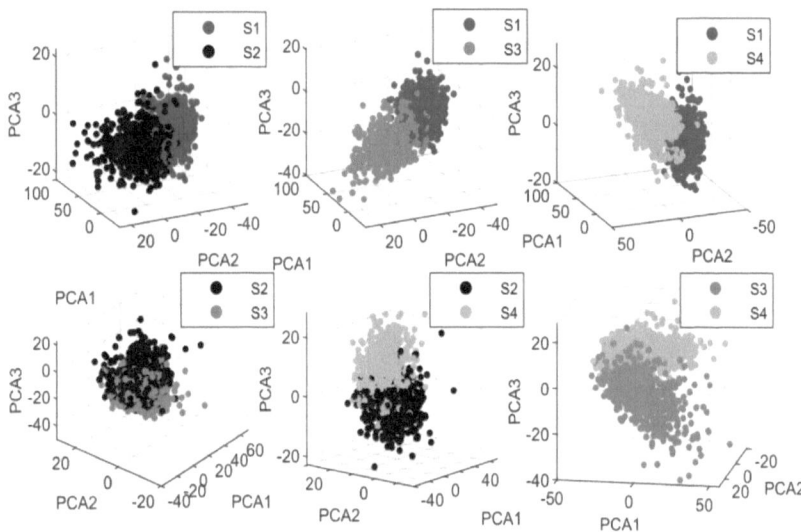

FIGURE 4.11 Examples of neuronal firing patterns analysed using an autoencoder. Each dot represents the firing pattern of a specific neuron. PCA was applied to results for better visualization.

test: S1–S3 vs S1–S4, $p = 0.049$; S1–S4 vs S2–S3, $p = 0.001$; S2–S3 vs S2–S4, $p = 0.030$; S2–S3 vs S3–S4, $p = 0.013$).

We next studied the topology structure and properties of neuronal population activity for spatial orientation. In comparison to non-spatial stimuli experiments, neuronal ensembles showed much more synchronized activity. Different topologies were observed in each spatial segment. Specifically, neurons in S1 and S4 showed considerably denser structure, and stronger influence on the network than neurons in S2 and S3 (Figure 4.12A–B). Additionally, neurons in S1 and S4 exhibited substantially higher clustering coefficients, whereas cell groups displayed more localized connectivity (Figure 4.12C–D). Additionally, the population activity was projected to a low-dimensional subspace by employing neural manifold analysis. Different population dynamics were observed in the four spatial segments, with large manifold amplitudes in S1 and S4 (Figure 4.13A).

Criticality measures remained stable during the first half and the second half of the experiment although the SC error reduced significantly over time, suggesting configuration closer to the critical point (SC, Student's t-test, $p = 0.0026$; DCC, Student's t-test, $p = 0.1768$; BR, Student's t-test, $p = 0.5686$; Figure 4.13B–C). However, it is worth noting that SC is a noisier metric than

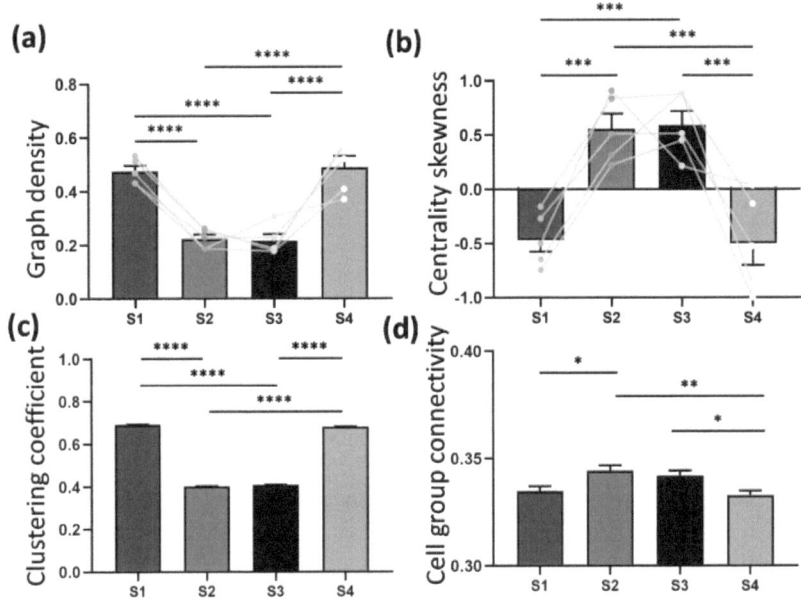

FIGURE 4.12 Neural network graph topology and neural manifolds in linear track experiments. (A) Graph density of the network in S1 and S4 is higher than that in S3 and S4 (one-way ANOVA, $F_{3,12} = 33.17$, $p < 0.0001$; post hoc test: S1–S2/S3, $p < 0.0001$; S4–S2/S3, $p < 0.0001$). (B) There are more neurons with stronger influence on the network in in S1 and S4 than S2 and S3 (one-way ANOVA, $F_{3,12} = 25.68$, $p < 0.0001$; post hoc test: S1–S2, $p = 0.003$; S1–S3, $p = 0.003$; S2–S4, $p = 0.003$; S3–S4, $p = 0.002$). (C) Neurons in S1 and S4 have stronger tendency to cluster together than S2 and S3 (one-way ANOVA, $F_{3,9717} = 8005$, $p < 0.0001$; post hoc test: S1–S2/S3, $p < 0.0001$; S4–S2/S3, $p < 0.0001$). (D) Neurons in S1 and S4 showed higher localized connectivity (one-way ANOVA, $F_{3,9717} = 6.206$, $p = 0.0164$; post hoc test: S1–S2, $p = 0.0001$; S2–S4, $p = 0.0014$; S3–S4, $p = 0.0199$).

others, precluding definite conclusions about the overall effect of these changes on system dynamics.

4.2.4 Neural Representation During Mixed Light–Sound Stimuli

As we had shown sensitivity to non-spatial stimuli, we next examined how mixed non-spatial stimuli affect hippocampal CA1 neuronal activity by

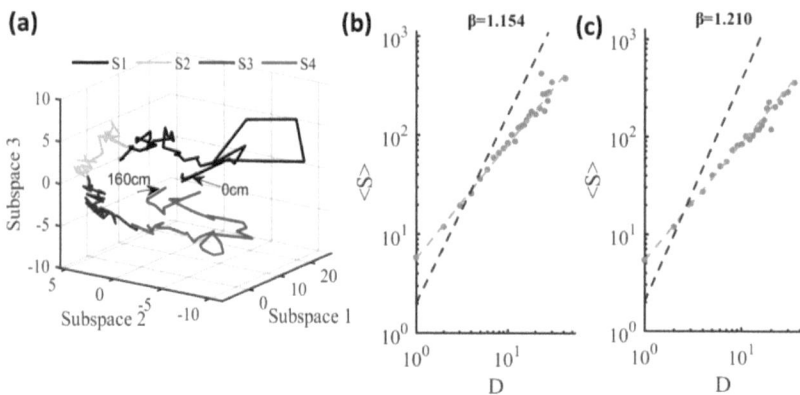

FIGURE 4.13 Neural manifolds representation and criticality analysis in linear track experiments. (A) An example of the neural manifolds representation. (B–C) Criticality analysis of the neuronal population in linear track experiments. Power–law distributions of the size (S), duration (D) in the (B) first half and (C) the second half of the experiment. The grey line represents the fitted β, and the red line represents the predicted β.

analysing neuronal population activity when mice ($n = 5$) were exposed to mixed light-sound stimuli. The signal was sectioned into four different types of epochs in four environments (DQ: dark + quiet; DS: dark + sound; LQ: light + quiet; LS: light + sound). At the beginning of each epoch, a substantial proportion of neurons responded instantly to stimuli, but the quantity in DQ ($35.88\% \pm 6.63\%$) and DS ($36.27\% \pm 8.19\%$) epochs was much higher than that in LQ ($17.77\% \pm 4.65\%$) and LS ($21.79\% \pm 4.87\%$) epochs (Figure 4.14A). We next examined coding properties of each neuron. Intriguingly, similar temporal information content and cosine similarity index were observed between DQ and DS epochs, LQ and LS epochs, respectively. The information content was at a relatively high level at the beginning of each epoch, then reduced rapidly (DQ, DS) or gradually (LQ, LS) to a similar level. On average, the information content was higher in LQ and LS epochs than that in DQ and DS epochs (Figure 4.14B). The temporal stability of neuronal firing patterns decreased in all scenarios with a relatively high cosine similarity index at the beginning. Contrary to the information content result, neurons in DQ and DS epochs revealed substantially higher stability (Figure 4.14C). Although there were no obvious changes in information content and stability of neural firing patterns, the autoencoder effectively split firing patterns into different clusters (Figure 4.15). Neurons showed different sensitivity to stimuli. In contrast to other scenarios, LQ–LS and DQ–DS had

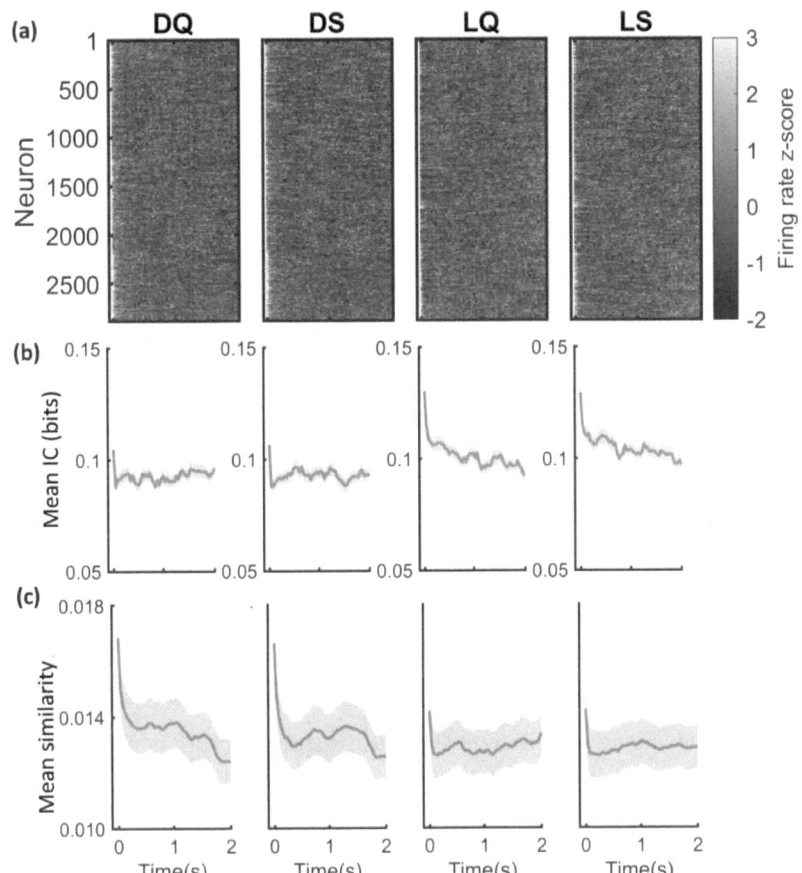

FIGURE 4.14 Visual and auditory mixed stimuli evoked distinct neuronal firing patterns in different windows. The brain activity was sectioned into four different types of epochs (DQ: dark + mute; DS: dark + sound; LQ: light + mute; LS: light + sound). (A) The average neuronal temporal firing rate map over trials when the animal was in different environments. (B) Neurons in LQ and LS epochs showed higher information content than that in DQ and DS epochs. (C) Neuronal firing patterns in DQ and DS epochs showed high stability across trials than that in LQ and LS epochs when measured using the cosine similarity index.

fewer reactive neurons, possibly implying intrinsically weaker sensitivity to sound.

We next studied the network topology under mixed stimuli (Figure 4.16A–D). Although the graph density and each neuron's impact on the network were

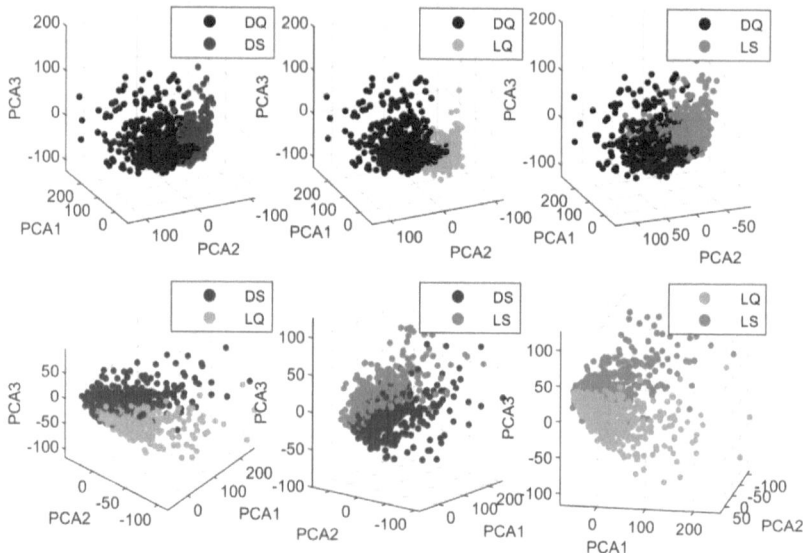

FIGURE 4.15 Examples of neuronal firing patterns analysed using an autoencoder. Each dot represents the firing pattern of a specific neuron. PCA was applied to the results for better visualization.

comparable among the different stimulus patterns, network connectivity was altered. The clustering coefficient of neuronal populations in LQ and LS epochs was substantially higher than that in DQ and DS epochs. Additionally, neurons in LS epochs displayed a more localized connectivity. The neural manifold representing population activity under different stimuli intertwined together after dimensionality reduction, with similar manifold amplitudes (Figure 4.17A). Criticality measures were estimated by diving the time epochs in half; the system again showed evidence of critical behaviour, but all metrics were stable over time (DCC, Student's t-test, $p = 0.814$; SC, Student's t-test, $p = 0.683$; BR, Student's t-test, $p = 0.904$; Figure 4.17B–C).

4.3 DISCUSSION

In this study, we recorded both spatial and non-spatial modality responses in the rodent hippocampus using widefield imaging of hundreds of neurons. The use of multi-neuronal ensemble recording during multisensory stimulation, in

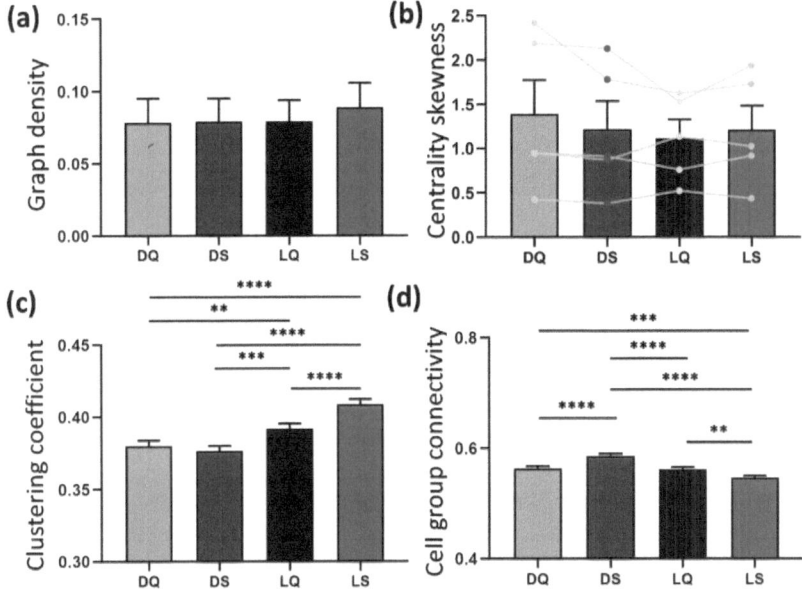

FIGURE 4.16 Neural network graph topology and neural manifolds in mixed stimuli experiments. (A) Neuron pairs in LS epochs have larger correlation coefficients than the others (one way ANOVA, $F_{3,196965} = 203.3$, $p < 0.0001$; *post hoc* test: LS-DQ/DS/LQ, $p < 0.0001$). (B) Network centrality distribution did not show significant differences (one-way ANOVA, $F_{3,12} = 1.465$, $p = 0.2733$). (C) Neurons in LS and LQ epochs have stronger tendency to cluster together than those in DS and DQ epochs (one-way ANOVA, $F_{3,8670} = 32.17$, $p < 0.0001$; *post hoc* test: LS-DQ/DS/LQ, $p < 0.0001$; LQ-DS, $p = 0.0002$; LQ-DQ, $p = 0.0081$). (D) Neurons in LS windows have more localized connectivity than others (one-way ANOVA, $F_{3,8670} = 28.46$, $p < 0.0001$; *post hoc* test: LS-DQ, $p = 0.0004$; LS-DS, $p < 0.0001$; LS-LQ, $p = 0.0022$; LQ-DS, $p < 0.0001$; DQ-DS, $p < 0.0001$).

combination with advanced network analysis, has yielded valuable insights into the intricacies of neuronal dynamics and coding patterns within hippocampal circuits, significantly advancing our understanding of the relational theory and cognitive maps theory. We demonstrate that the hippocampus processes non-spatial information independently of spatial inputs. The network's topological configuration demonstrates dense interconnections for spatial information while exhibiting sparse connectivity for non-spatial information. Furthermore, neuronal populations exhibit distinct coding mechanisms depending on sensory inputs. These findings contribute to our understanding of hippocampal neuronal population activity across different information modalities and offer potential support for the relational theory.

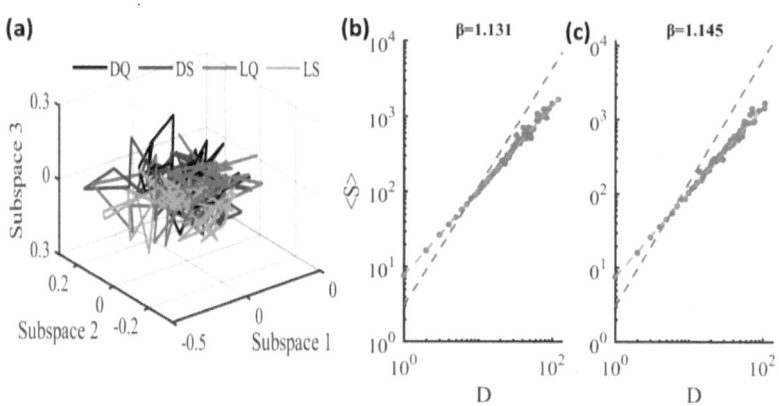

FIGURE 4.17 Neural manifolds representation and criticality analysis in mixed stimuli experiments. (A) An example of the neural manifolds representation in mixed stimuli experiments. (B–C) Criticality analysis of the neuronal population in mixed stimuli experiments. Power–law distributions of the size (S), duration (D) in the (B) first half and (C) the second half of the experiment. The grey line represents the fitted β, and the red line represents the predicted β.

The hippocampus is a key structure in the brain that processes spatial information. It is not generally considered a part of the visual or auditory system. However, several pathways for both are well described. Visual inputs are transmitted to the hippocampus through the surrounding neocortex via multisynaptic pathways[13,14]. Auditory inputs project from polysensory cortex[15], then to the hippocampus through a medial septum relay[16,17]. Thus, visual and auditory signalling to the hippocampus likely undergoes distinct integrative stages that affect neural activity propagation. In contrast to previous work[18-20], where neuronal response latencies to visual or auditory stimuli exhibited irregular random distributions, the majority of neurons showed rapid responses in our experiments. This is possibly due to different experimental design as visual or auditory stimuli are coupled with other sensory modalities in previous experiments. For example, previous study recorded responses on a rotating plate and animal movement might have influenced the neural activity[19].

By assessing the information content and cosine similarity index of neuronal firing patterns, we found that CA1 neurons displayed different coding complexity and stability depending on the sensory inputs. Visual and auditory stimuli modulated temporal features of neural activity differentially. As opposed to auditory information, which was coded over a long period of time, visual information was primarily encoded as stimuli occurred. Additionally, neuronal activity with high information content typically displayed high stability in visual experiments but low stability in auditory experiments. In contrast to non-spatial modalities, neural activity was strongly modulated by spatial inputs via more complex but stable processes. These differences may be ascribed to the intrinsic properties of hippocampal neurons or due to different information flow pathways. It will be of considerable interest to study the neuronal activity in the major hippocampal input entorhinal cortex, which is increasingly believed to have a perceptual function[21].

Although some deep neural network-based classifiers allow the classification of patterns with great accuracy, these algorithms require extremely large datasets and are challenging to determine which specific features inform the categorization decision[22]. Instead of constructing a classifier, we used an unsupervised autoencoder to extract internal features of the patterns. In all scenarios, patterns could be distinguished even in conditions where the information content and cosine similarity index showed comparable results. It is worth noting that the patterns in both spatial and non-spatial experiments can be described using around 10 features. As multisensory experiences are consolidated in the hippocampus[23], this number may imply intrinsic characteristics of the hippocampal network, like the number of senses. Future model work or BCI applications may benefit from this finding.

Application of graph theory provides a coherent approach to study functional connectivity in a regional network or globally[24]. Previous studies pointed out that the hippocampal network in spatial cognition was topological[25], which was supported by the results of current experiments. Additionally, we found that hippocampal network topology dynamically reconfigured under different conditions in response to various stimuli. Such reconfiguration may form a fundamental neurophysiological mechanism for cognitive functions as suggested by previous studies[26].

Recent findings that rhythmic light flicker can rescue hippocampal low gamma oscillation and improves cognition in Alzheimer's disease mouse models[11] suggest an important role of rhythmic stimuli in tuning hippocampal neural activity and the network restructuring observed in the current study may potentially be related to this observation. It would be of considerable interest to directly observe the effects of these flicker stimuli on network structure and criticality, as we have recently correlated criticality with cognitive tasks using the same widefield calcium imaging technique.

It has been hypothesized that population activity in cortical regions is confined in neural manifolds, which are low-dimensional parameter spaces embedded in the original neural state space[27]. We observed distinct temporal feature patterns for different types of information, indicating distinct information processing mechanisms. In visual stimuli experiments and linear track experiments, encoding geometries varied smoothly in time, but somewhat abruptly in auditory stimuli experiments. An intriguing finding is that manifold activity under different stimuli intertwined in mixed stimulus experiments, suggesting similar information processing mechanisms in this case, possibly due to the complexity of the stimuli, which might overload the animal's cognitive capacity. Previous study suggests that local regions of neural manifolds could have further substructures related to conditions[28]. This hypothesis was validated in our experiments, where different frequencies occupied separate regions in the corresponding neural manifolds. However, the population activity in mixed stimulus experiments showed certain overlap in neural manifolds, showing that further studies are needed to define manifold complexity and structure.

Recent studies have attempted to show that neural ensembles display self-organized criticality, especially in the context of potential functional efficiency that may accrue from such regimes[29-31]. Brain criticality analysis has been applied to a variety of clinically relevant domains such as epilepsy, neurodegenerative disease, anaesthesia, and psychiatry[32]. It has been suggested as an efficient measurement to study the information processing[33]. Measuring several metrics that are necessary conditions for critical behaviour, we found that indeed hippocampal CA1 neuronal ensembles appeared to display critical dynamics, but with varying properties depending on the type of stimuli observed. DCC, SC, and BR are commonly used measurements to quantify the system criticality. The measurements of DCC and SC are both calculated based on power–law distributions of avalanches properties. In contrast, the BR approach measures the percentage of triggered events among all the events and does not employ the avalanches construct. However, whether branching process like models can reproduce the statistics of spike avalanches remains controversial[34]. In visual and auditory stimuli experiments, system dynamics moved away from the critical point as time passed considering increased DCC errors. While BR errors only increased significantly in visual but not in auditory experiments, showing that changes in avalanches properties did not always lead to alterations in branching processes. Compared with visual and auditory stimuli experiments, the system dynamics did not change in mixed stimuli experiments. This could be attributed to the need for efficient information processing as the stimuli became more complicated in these experiments.

To ensure a substantial number of trial recordings for our analysis, we conducted long-time recordings. However, a significant challenge we encountered was the photobleaching of fluorescent calcium indicators. Consequently, we were able to record 300 trials in visual stimulus experiments, but only 100 trials/frequency in auditory stimulus experiments. Given the substantial difference bias in trial numbers and the variations in stimulus repeat patterns, it may not be suitable to directly compare the results between these two sets of experiments. To address this issue, we conducted a mixed stimuli experiment to compare the responses elicited by visual and auditory stimuli. In this scenario, we had four different background environments. To mitigate photobleaching, we needed to shorten the duration of each trial. Therefore, direct result comparisons with the other two experiments may not be appropriate due to the differences in stimulus patterns. To tackle the photobleaching issue and facilitate result comparisons in future studies, one method is to utilize UCLA miniscope version 4. This updated version incorporates a more sensitive Python CMOS chip and could substantially extend the recording length.

In summary, this study advances our understanding of the hippocampal population activity in response to diverse modalities of information and offers innovative perspectives on the interpretation of relational theory and cognitive maps theory. Our findings demonstrate that while the CA1 neuronal population primarily processes spatial information, it also exhibits a considerable degree of sensitivity to non-spatial modalities that are independent of spatial inputs. These results provide support for the relational theories.

4.3.1 Limitations of Study

While some cells exhibit non-spatial responses independent of spatial ones, others may display spatial responses independently of non-spatial ones, and some may exhibit combined spatial and non-spatial responses. Thus, the current study only provides support for the relational theory, and we cannot definitively conclude that the cognitive map theory is incorrect. One possible avenue for future research is to explore this by deploying an infrared eye tracking system to accurately detect the animal's spatial input in the chamber.

4.4 METHODS

4.4.1 Experimental Model and Subject Details

All procedures were carried out in accordance with the Australian Animal Welfare Committee guidelines and approved by Florey Animal Ethics Committee (No. 18-008UM). Male C57BL/6 mice (aged 10–12 weeks, 23–25 g) were maintained on a 12-hour light/dark cycle with water and standard mice chow *ad libitum*.

4.4.2 Surgery Protocols

We first injected 500 nL of pAAV.Syn.GCaMP6f.WPRE.SV40 virus (viral load: 2.2×10^{13} GC/mL; AddGene, USA) into dorsal hippocampus (coordinates: AP -2.1 mm, ML +2.1 mm, DV -1.7 mm). The injection speed was less than 35 nL/min and the injector was left in place for extra 10 min to allow for viral diffusion. One week after viral transduction, a round craniotomy (diameter: 2 mm) was made at -2.1 mm posterior and 1.6 mm right of bregma. Overlying brain tissue was aspirated to expose hippocampal vertical striations with a 27-gauge blunt needle and artificial cerebrospinal fluid was applied to provide a clear operating field. A GRIN lens (0.23 pitch, No. 64-519, Edmund Optics) was lowered to the bottom of the craniotomy (-1.35 mm from the top of the skull) and secured with cyanoacrylate. Two M1 anchor screws were fixed close to the lens (coordinates: AP +1.8, ML -2.5; AP -2.8, ML -0.8) and dental acrylic was built around the lens for support. Carprofen (5 mg/kg), dexamethasone (0.5 mg/kg), and enrofloxacin water were injected after the surgery and provided to the animal daily for one week. Five weeks following the implantation surgery, neurons were observed using a miniscope with a baseplate attached at the bottom. Upon finding an optimal position for imaging, the baseplate was mounted on the head cap with dental acrylic, and the focal length adjustment screw was tightened in place. Throughout the experiment, we encountered challenges such as animal mortality, headpiece detachment, and signal loss. Thus, different mice were used in each experiment.

4.4.3 Experimental Protocol

On the day of the experiment, the experimental animal was brought into a silent room 30 min prior to the beginning of the recording to acclimate to the new environment and the miniscope was attached with optimized LED excitation intensity.

In the linear track experiment, trained mice traversed a 1.6 m linear track for 16 running trials, and neural activity recorded. A camera fixed overhead was synchronized with the miniscope to track the animal's location. In light stimulus experiments, an LED light source was mounted above the chamber and the luminance was set at 380 lux. Previous work[19] found that neuronal populations exhibited widely distributed response latencies to stimuli, with a maximum around 2 s. Consequently, we selected a 2-s stimulus duration and employed a 3-s interval to minimize any significant influence of the previous epoch's stimulus on the current epoch's neuronal responses. The light source was activated (2 s) and deactivated (3–3.5 s) alternately 300 times. With sound stimuli, a speaker was fixed on top of the chamber and the intensity was set at 70 dB SPL, providing three pure sinusoidal tones centred at 4, 8, and 16 kHz, lying within both human and mouse hearing sensitivity ranges and including the most sensitive frequencies in the mouse audiogram[35]. The auditory stimuli were activated (2 s) and deactivated (3–3.5 s) alternately in a random order with the stimulus at each frequency repeated 100 times. With light-sound mixed stimuli, the visual and auditory stimuli (pure sinusoidal tone centred at 8KHz) were combined. In total, four different background environments (dark + silent; dark + sound; light + silent; light + sound) in a random order were used. Each environment was 2 s in duration and repeated 200 times. The light source and speaker were controlled and synchronized with the miniscope by a Raspberry Pi 3 board with custom-designed software.

4.4.4 Calcium Activity Analysis

The analysis pipelines were similar to our previous reports[36]. Briefly, a non-rigid motion correction algorithm was applied first to register the raw images[37]. Next, a constrained non-Negative Matrix Factorization for Endoscopic (CNMF-E) recordings algorithm was used to extract neural calcium activity[38]. The calcium activity of each neuron was deconvolved using a pool adjacent violators algorithm[39]. Finally, the temporal neural activity was binarized to obtain the temporal calcium events[40]. In the linear track experiment, the data on each end of the track was excluded from the analysis.

4.4.5 Information Content

Information content[41] measures the coding complexity of each neuron, and a larger value indicates higher complexity. In the linear track experiment, the data were divided into 1 cm spatial bins for analysis. In stimuli experiments, the binning method was converted from spatial domain to temporal domain (bin=1/Fs, where the sampling frequency Fs=30 frame/s). The information content is defined as:

$$I = \sum_{i=1}^{K} P_i \frac{\lambda_i}{\bar{\lambda}} \log_2 \frac{\lambda_i}{\bar{\lambda}}, \bar{\lambda} = \sum_{i=1}^{K} P_i \lambda_i \qquad (4.1)$$

where K represents the number of spatial bins (linear track experiment) or the length of the epoch data (stimuli experiments), P_i is the occupancy ratio of the

ith bin, λ_i is the neural calcium event rate in the ith bin, and $\bar{\lambda}$ is the mean calcium event rate.

4.4.6 Cosine Similarity Index

A "cosine similarity index" between individual neurons was used to quantify the stability of neuronal firing patterns. The similarity of each neuron was defined as the normalized inner product of the activity between every two running trials or stimuli epochs[42],

$$S = \frac{1}{N(N-1)} \sum_{i=1}^{N-1} \sum_{j=i+1}^{N} \frac{2 \cdot C_i C_j}{\left(C_i^2 + C_j^2\right)} \tag{4.2}$$

where S represents cosine similarity index, N represents the total number of running trials or stimuli epochs, C_i and C_j are neural calcium event vectors in the ith and jth *trials* or epochs. This metric is restricted to [0,1], where "1" means the neuronal firing patterns are the same in all trials or epochs, while "0" means they are completely different.

4.4.7 Neuronal Firing Pattern Extraction

The information content and cosine similarity index are two metrics that quantify differences in neural firing patterns, but they may fail to detect non-obvious differences in some cases (see Results). We used an autoencoder to detect differences in neuronal firing patterns under different contexts in a high-dimensional parameter space. An autoencoder is a type of unsupervised artificial neural network that learns to copy its input to its output through layers with fewer neurons, providing useful dimensionality reduction of complex data spaces[43]. It combines two parts: encoder and decoder. The encoder decreases the dimensionality of data, and the decoder learns to reconstruct the original input. During training, the encoder identifies the most significant components from the original data. We constructed an autoencoder using the Tensorflow machine learning platform[44] to decrease the dimensionality of the neural calcium events epoch data. The inputs of the model were the spatial firing rates of each neuron in the running trials or temporal firing rates in each epoch. Both the encoder and decoder were constructed with three dense layers and SELU activation functions. A binary cross-entropy loss function and an Adam optimizer were chosen to train the model and a grid search method fitted the parameters including learning rate, the number of nodes in each layer, and the number of training epochs to minimize the reconstructed data error[36]. Given the substantial workload involved in optimizing parameters using datasets from five animals in four different experiments, we chose not to conduct a comprehensive grid search for hyperparameter optimization. We began by identifying and tuning the most impactful hyperparameters while keeping others at reasonable values. The testing values for the grid search were determined based on previous experience and some random testing. The mean reconstruction error was found to be 3.84% ± 1.60% in visual stimuli experiments, 4.91% ± 1.87% in auditory stimuli experiments, 5.37% ± 1.93% in linear track experiments, and 5.28% ± 1.85% in mixed stimuli experiments. In the encoder, we found that a configuration of 1200 nodes in the first dense layer, 120 nodes in the second dense layer, and 15 nodes in the third dense layer was a typical good choice for achieving satisfactory reconstruction. In the decoder, the number of nodes in each dense layer is reversed, with 15 nodes in the first dense layer, 120 in the second, and 1200 in the third. As these features were described in a high-dimensional space (>3 dimensions), they could not be easily visualized. Consequently, we performed PCA solely for visualization purposes. Following the feature extraction with the autoencoder, we calculated the cluster distance based on these high-dimensional features. This analysis aimed to assess whether firing patterns under different scenarios could be effectively separated. The distance of two clusters was defined as L2-Norm of centroids calculated using a K-means clustering algorithm.

4.4.8 Weighted Normalized Mutual Information

"Mutual information" is a measure of mutual dependence between variables. WNMI considers the weights of the deconvolved neural activity and adds a normalization term, which is used to detect reactive neurons in response to the various stimulus contexts. Mutual information yields values from 0 to +∞ and may introduce different levels of bias depending on the datasets. Normalization restricts its values to the range [0, 1] and compensates for the bias towards multivalued features[45]. The amplitude of calcium events deconvolved from analysis pipelines can be interpreted as the probability that a neuron fires at each frame.

WNMI is defined as:

$$I_{WN}(X,Y) = \frac{I_W(X,Y)}{\min\left(U_{WX}(X,Y), U_{WY}(X,Y)\right)} \tag{4.3}$$

$$I_W(X,Y) = \sum_{y \in Y}\sum_{x \in X} W(y)P(x,y)log_2\left[\frac{P(x,y)}{P(x)P(y)}\right] \tag{4.4}$$

$$W(y) = \frac{1}{1+e^{-[f(y)-u]}} \tag{4.5}$$

$$U_{WX}(X,Y) = \sum_{y \in Y}\sum_{x \in X} W(y)P(x,y)\left\{-log_2\left[P(x)\right]-1+\frac{P(x)P(y)}{P(x,y)}\right\} \tag{4.6}$$

$$U_{WY}(X,Y) = \sum_{y \in Y}\sum_{x \in X} W(y)P(x,y)\left\{-log_2\left[P(y)\right]-1+\frac{P(x)P(y)}{P(x,y)}\right\} \tag{4.7}$$

where X represents different stimuli scenarios or spatial locations, Y represents firing rates of the calcium events, $I_{WN}(X,Y)$ and $I_W(X,Y)$ are the WNMI and weighted mutual information between X and Y, respectively, and $U_{WX}(X,Y)$ and $U_{WY}(X,Y)$ are normalization terms that represent weighted entropy of variables X and Y. $W(y)$ is the weighting component that takes the form of a logistic function, where $f(y)$ is the amplitude of the neural activity at each time frame that represents the neuronal firing probability and u is the mean amplitude. $P(x,y)$, $P(x)$, and $P(y)$ represent the joint probability density function and probability density functions of X and Y, respectively.

4.4.9 Neuronal Sensitivity to Stimuli

We first measured the WNMI when comparing the activity between every two stimuli or spatial locations for each neuron. Then, the temporal or spatial neural calcium events were shuffled 500 times and the average WNMI was evaluated for each shuffle. For a neuron to be classified as a reactive, the average WNMI must be above chance ($p<0.05$) with respect to the permutation results[18,40].

4.4.10 Network Graphs

We quantified patterns of neural population activity of different experimental scenarios using network graphs. This analysis detects non-obvious patterns of neuronal ensemble topological and dynamical responses to various forms of perturbation. We measured the pairwise Pearson correlation between all neuronal activities and set the absolute correlation value (threshold: 0.05) as the edge weight between two neurons in the network graph. One issue with this method is that it can yield negative correlation values, which can be challenging to interpret. In previous studies, two methods have been employed to handle negative values. One approach involves using absolute values, while the other entails setting negative weights to zero[46]. Both methods have their limitations. The former focuses only on the magnitude of the negative weights, while the latter considers only their sign. In our experiment, we observed a substantial number of negative correlation values, leading us to take the absolute value of correlation coefficients just before applying thresholding. Regarding the choice of the correlation value threshold, there is no consensus within the scientific community on how to select specific thresholds. Using a lower threshold has the advantage of retaining all information, which can be important in certain applications. However, this may result in the inclusion of many noisy or irrelevant connections in the analysis, potentially diluting meaningful relationships. One approach to partially address this issue is to use the correlation value as the edge weight. Assigning weighted edges provides a means to assess the importance of connections within the graph, enabling a more nuanced analysis. In our analysis, we used weighed edges in the graph and opted for a relatively low threshold of 0.05. We refrained from using lower values due to the extremely extended Markov diffusion time. This parameter is directly proportional to the number of random walks required for a walker to diffuse out of a cell group. We measured several fundamental network properties, graph density, clustering coefficient, cell group connectivity, and eigenvector centrality.

The graph density is the proportion of edges present in a graph divided by the maximum possible edges and indicates the ratio of possible relationships in the network that are actually present. The clustering coefficient measures the degree to which nodes in a graph tend to cluster together. It is defined as the number of triangles in the network divided by the number of connected triples of vertices[47]. Cell groups are detected by evaluating modularity, a measure that quantifies the network's division strength[48]. The cell group connectivity measures the connectivity of different cell groups. It is defined as the number of edges a node connects with its own cell group divided by all edges it connects with other neurons, and a high value indicates more localized connectivity. Eigenvector centrality[49] measures the transitive influence of a node in a network while taking into account the influence of neighbours. A node with few connections can have a very high eigenvector centrality score if those connections are to very well-connected others. It is defined as:

$$c\left(v_i\right) = \sum_{j=1}^{n} a_{i,j}\, c\left(v_j\right),\ Ac = \lambda c \qquad (4.8)$$

where n is the number of neighbours of the node v_i, $A = a_{i,j}$ is the adjacency matrix generated by measuring the pairwise Pearson correlation, and λ is the eigenvalue. The centrality scores are normalized such that the sum of all scores equals to 1.

4.4.11 Neural Manifolds

Several cortical neuronal populations share fundamental mechanisms or principles governing cortical processing, showing similar patterns of activity. Geometrical analysis of neural population activity has led to the identification of topologically defined state spaces termed "neural manifolds" with reduced dimensionality. It is thought that manifolds may better identify dynamically and computationally significant functional structures[50]. To project the population activity from a high-dimensional neural state space to low-dimensional neural manifolds, we employed an autoencoder for dimensionality reduction. The structure of the autoencoder was similar to the one used in neuronal firing pattern extraction mentioned above with the same optimization method, but the inputs of the model were the neural activity of all neurons at each time point. The outputs from the encoder defined the features/subspaces describing the manifold activity and were used for manifold analysis. The extracted features indicated whether the neuronal population adapts its neural representations in response to external information, such as stimuli. If the population activity at two different time points has similar feature values, this means that the network represents similar information, and vice versa. Furthermore, these features can demonstrate differences in information processing mechanisms for various types of information. We also calculated the Euclidean norm of each temporal vector and the Euclidean angle between each vector with respect to the first temporal vector in manifolds. Manifold amplitudes and angles provide two ways to quantify the similarity between neural modes and indicating when self-similarity changes. This angular analysis is a valuable tool to observe shifts in neural activity patterns over time. When the angle between two vectors is large, it indicates dissimilar population activity patterns at those two time points.

4.4.12 Criticality Analysis

Criticality denotes a mode of population activity between states of order and chaos, and it has been proposed that neural population activity with critical dynamics are optimized for information processing and memory capacity. It is still unclear how to determine whether a system is in a critical state or not. While meeting the power–law distribution is a basic requirement, recent literature suggests testing additional metrics[33]. We separately evaluated and compared dynamic states of population activity in the first half and the second half of the different experimental scenarios using four metrics: (1) power–law distributions of the size (S) and duration (D) of neuronal avalanches; (2) DCC; (3) SC error; (4) BR.

Recent studies have revealed that neuronal populations can generate activity avalanches (a cascade of bursts of activity) with size and duration described by a power–law distribution [51,52]. We first obtained the "population activity" as the sum of all detected calcium events within each time bin. The neuronal avalanches were detected using a threshold method (>60% of median population activity)[53], where an avalanche began when the population activity crossed the threshold from below and terminated when it reached the threshold from above. Then, the avalanche size (number of total calcium events) and duration (number of time bins) distributions were fitted with a truncated power–law using a maximum-likelihood estimation method[54]:

$$f(x) = A\left(\mu, x_{min}, x_{max}\right)x^{-\mu}, \ A\left(\mu, x_{min}, x_{max}\right) = \frac{1}{\sum_{x=x_{min}}^{x_{max}} x^{-\mu}} \tag{4.9}$$

where x represents avalanche size or duration; $f(x)$ is the probability mass function; A is a normalized constant; μ represents the power–law exponent τ (avalanche size) or α (avalanche duration).

The DCC is a measure of how close a system operates to criticality[55]. As avalanche size scales with the average duration in a critical system according to $\langle S \rangle = D^\beta$, DCC is defined as the absolute difference between estimated $\beta = (\alpha - 1)/(\tau - 1)$ and fitted β, and a smaller DCC value the closer the system is to criticality[52].

When a neural system operates in a critical regime, the profiles of avalanches with different durations should have the same scaled mean shape[52]:

$$s(t, D) \propto D^\gamma F\left(\frac{t}{D}\right), F\left(\frac{t}{D}\right) = \left\langle \frac{s(t, D)}{D^\gamma} \right\rangle \tag{4.10}$$

where $s(t, D)$ represents the number of calcium events at time t given avalanche duration of D; $\gamma = \beta - 1$; $F\left(\dfrac{t}{D}\right)$ represents avalanche temporal profiles. A collection of temporal profiles $F\left(\dfrac{t}{D}\right)$ can be extracted given different duration D, and the proximity to a predicted scaled mean shape can be measured by the SC error, which is calculated as $\mathrm{var}(F)/(\max(F)-\min(F))^2$.

The BR quantifies the proportion of events at time $t+1$ to those at time t, and a critical system is expected to have a BR equal to 1. We utilized a multiple linear regression-based method[56] to get an unbiased estimation of the BR:

$$< N(t+1)|N(t)\rangle = mN(t)+h \tag{4.11}$$

$$< n(t)|N(t)\rangle = \mu N(t)+\varepsilon \tag{4.12}$$

$$m^k = m_k \,/\, \frac{\mu^2 \mathrm{var}(N(t))}{\mathrm{var}(n(t))} \tag{4.13}$$

where $N(t)$ is the real activity and $n(t)$ is the subsampled activity; m is the BR; h represents external stimuli; μ and ε are constants; m_k is the multiple linear regression coefficient of the subsampled activity between time t and $t + k$.

4.4.13 Quantification and Statistical Analysis

Statistical analyses were performed using GraphPad Prism software. The level of significance was set at $p < 0.05$. Statistical significance was assessed by two-tailed paired Student's t-test or one-way repeated-measures ANOVA with Bonferroni *post hoc* comparisons. Normality of the data was confirmed by Kolmogorov–Smirnov and Shapiro–Wilks test. All results are shown as mean ± SEM.

REFERENCES

1. Eichenbaum, H., Dudchenko, P., Wood, E., Shapiro, M., and Tanila, H. (1999). The hippocampus, memory, and place cells: is it spatial memory or a memory space? *Neuron* 23, 209–226.
2. Bird, C.M., and Burgess, N. (2008). The hippocampus and memory: insights from spatial processing. *Nature Reviews Neuroscience* 9, 182–194.
3. Lisman, J., Buzsáki, G., Eichenbaum, H., Nadel, L., Ranganath, C., and Redish, A.D. (2017). Viewpoints: how the hippocampus contributes to memory, navigation and cognition. *Nature Neuroscience* 20, 1434–1447.

4. O'keefe, J., and Nadel, L. (1979). Précis of O'Keefe & Nadel's The hippocampus as a cognitive map. *Behavioral and Brain Sciences* 2, 487–494.
5. O'Keefe, J., and Krupic, J. (2021). Do hippocampal pyramidal cells respond to nonspatial stimuli? *Physiological Reviews* 101, 1427–1456.
6. Aronov, D., Nevers, R., and Tank, D.W. (2017). Mapping of a non-spatial dimension by the hippocampal–entorhinal circuit. *Nature* 543, 719–722.
7. Behrens, T.E., Muller, T.H., Whittington, J.C., Mark, S., Baram, A.B., Stachenfeld, K.L., and Kurth-Nelson, Z. (2018). What is a cognitive map? Organizing knowledge for flexible behavior. *Neuron* 100, 490–509.
8. Montijn, J.S., Meijer, G.T., Lansink, C.S., and Pennartz, C.M. (2016). Population-level neural codes are robust to single-neuron variability from a multidimensional coding perspective. *Cell Reports* 16, 2486–2498.
9. Stringer, C., Pachitariu, M., Steinmetz, N., Reddy, C.B., Carandini, M., and Harris, K.D. (2019). Spontaneous behaviors drive multidimensional, brainwide activity. *Science* 364, eaav7893.
10. Scharwächter, L., Schmitt, F.J., Pallast, N., Fink, G.R., and Aswendt, M. (2022). Network analysis of neuroimaging in mice. *Neuroimage* 253, 119110.
11. Zheng, L., Yu, M., Lin, R., Wang, Y., Zhuo, Z., Cheng, N., Wang, M., Tang, Y., Wang, L., and Hou, S.-T. (2020). Rhythmic light flicker rescues hippocampal low gamma and protects ischemic neurons by enhancing presynaptic plasticity. *Nature Communications* 11, 3012.
12. Ghosh, K.K., Burns, L.D., Cocker, E.D., Nimmerjahn, A., Ziv, Y., Gamal, A.E., and Schnitzer, M.J. (2011). Miniaturized integration of a fluorescence microscope. *Nature Methods* 8, 871–878.
13. Lavenex, P., and Amaral, D.G. (2000). Hippocampal-neocortical interaction: A hierarchy of associativity. *Hippocampus* 10, 420–430.
14. Ranganath, C., and Ritchey, M. (2012). Two cortical systems for memory-guided behaviour. *Nature Reviews Neuroscience* 13, 713–726.
15. Zhang, L., Wang, J., Sun, H., Feng, G., and Gao, Z. (2022). Interactions between the hippocampus and the auditory pathway. *Neurobiology of Learning and Memory* 189, 107589.
16. Vinogradova, O. (1975). Functional organization of the limbic system in the process of registration of information: facts and hypotheses. In *The Hippocampus: Volume 2: Neurophysiology and Behavior* (pp. 3–69). Boston, MA: Springer.
17. Kaifosh, P., Lovett-Barron, M., Turi, G.F., Reardon, T.R., and Losonczy, A. (2013). Septo-hippocampal GABAergic signaling across multiple modalities in awake mice. *Nature Neuroscience* 16, 1182–1184.
18. Itskov, P.M., Vinnik, E., Honey, C., Schnupp, J., and Diamond, M.E. (2012). Sound sensitivity of neurons in rat hippocampus during performance of a sound-guided task. *Journal of Neurophysiology* 107, 1822–1834.
19. Liu, Y.z., Wang, Y., Tang, W., Zhu, J.y., and Wang, Z. (2018). NMDA receptor-gated visual responses in hippocampal CA1 neurons. *The Journal of Physiology* 596, 1965–1979.
20. Xiao, C., Liu, Y., Xu, J., Gan, X., and Xiao, Z. (2018). Septal and hippocampal neurons contribute to auditory relay and fear conditioning. *Frontiers in Cellular Neuroscience* 12, 102.

21. Baxter, M.G. (2009). Involvement of medial temporal lobe structures in memory and perception. *Neuron* 61, 667–677.
22. Zhang, Y., Tiňo, P., Leonardis, A., and Tang, K. (2021). A survey on neural network interpretability. *IEEE Transactions on Emerging Topics in Computational Intelligence* 5, 726–742.
23. Fortin, N.J., Agster, K.L., and Eichenbaum, H.B. (2002). Critical role of the hippocampus in memory for sequences of events. *Nature Neuroscience* 5, 458–462.
24. Gonzalez, W.G., Zhang, H., Harutyunyan, A., and Lois, C. (2019). Persistence of neuronal representations through time and damage in the hippocampus. *Science* 365, 821–825.
25. Yuri, D., Brandt, V.L., and Frank, L.M. (2014). Reconceiving the hippocampal map as a topological template. *eLife* 3.
26. Braun, U., Schäfer, A., Walter, H., Erk, S., Romanczuk-Seiferth, N., Haddad, L., Schweiger, J.I., Grimm, O., Heinz, A., and Tost, H. (2015). Dynamic reconfiguration of frontal brain networks during executive cognition in humans. *Proceedings of the National Academy of Sciences* 112, 11678–11683.
27. Ebitz, R.B., and Hayden, B.Y. (2021). The population doctrine in cognitive neuroscience. *Neuron* 109, 3055–3068.
28. Koay, S.A., Charles, A.S., Thiberge, S.Y., Brody, C.D., and Tank, D.W. (2022). Sequential and efficient neural-population coding of complex task information. *Neuron* 110, 328–349. e311.
29. De Arcangelis, L., Perrone-Capano, C., and Herrmann, H.J. (2006). Self-organized criticality model for brain plasticity. *Physical Review Letters* 96, 028107.
30. Chialvo, D.R. (2010). Emergent complex neural dynamics. *Nature Physics* 6, 744–750.
31. Cocchi, L., Gollo, L.L., Zalesky, A., and Breakspear, M. (2017). Criticality in the brain: A synthesis of neurobiology, models and cognition. *Progress in Neurobiology* 158, 132–152.
32. Zimmern, V. (2020). Why brain criticality is clinically relevant: a scoping review. *Frontiers in Neural Circuits* 14, 54.
33. O'Byrne, J., and Jerbi, K. (2022). How critical is brain criticality? *Trends in Neurosciences*, 45(11), pp.820–837.
34. Ribeiro, T.L., Ribeiro, S., Belchior, H., Caixeta, F., and Copelli, M. (2014). Undersampled critical branching processes on small-world and random networks fail to reproduce the statistics of spike avalanches. *PLoS One* 9, e94992.
35. Ehret, G., and Riecke, S. (2002). Mice and humans perceive multiharmonic communication sounds in the same way. *Proceedings of the National Academy of Sciences* 99, 479–482.
36. Sun, D., Unnithan, R.R., and French, C. (2021). Scopolamine impairs spatial information recorded with "miniscope" calcium imaging in hippocampal place cells. *Frontiers in Neuroscience* 15, 640350.
37. Pnevmatikakis, E.A., and Giovannucci, A. (2017). NoRMCorre: An online algorithm for piecewise rigid motion correction of calcium imaging data. *Journal of Neuroscience Methods* 291, 83–94.

38. Zhou, P., Resendez, S.L., Rodriguez-Romaguera, J., Jimenez, J.C., Neufeld, S.Q., Giovannucci, A., Friedrich, J., Pnevmatikakis, E.A., Stuber, G.D., and Hen, R. (2018). Efficient and accurate extraction of in vivo calcium signals from microendoscopic video data. *elife* 7, e28728.
39. Friedrich, J., Zhou, P., and Paninski, L. (2017). Fast online deconvolution of calcium imaging data. *PLoS Computational Biology* 13, e1005423.
40. Shuman, T., Aharoni, D., Cai, D.J., Lee, C.R., Chavlis, S., Page-Harley, L., Vetere, L.M., Feng, Y., Yang, C.Y., and Mollinedo-Gajate, I. (2020). Breakdown of spatial coding and interneuron synchronization in epileptic mice. *Nature Neuroscience* 23, 229–238.
41. Ravassard, P., Kees, A., Willers, B., Ho, D., Aharoni, D., Cushman, J., Aghajan, Z.M., and Mehta, M.R. (2013). Multisensory control of hippocampal spatiotemporal selectivity. *Science* 340, 1342–1346.
42. Carrillo-Reid, L., Miller, J.-e.K., Hamm, J.P., Jackson, J., and Yuste, R. (2015). Endogenous sequential cortical activity evoked by visual stimuli. *Journal of Neuroscience* 35, 8813–8828.
43. Kramer, M.A. (1991). Nonlinear principal component analysis using autoassociative neural networks. *AIChE Journal* 37, 233–243.
44. Pang, B., Nijkamp, E., and Wu, Y.N. (2020). Deep learning with tensorflow: A review. *Journal of Educational and Behavioral Statistics* 45, 227–248.
45. Estévez, P.A., Tesmer, M., Perez, C.A., and Zurada, J.M. (2009). Normalized mutual information feature selection. *IEEE Transactions on Neural Networks* 20, 189–201.
46. Kruschwitz, J., List, D., Waller, L., Rubinov, M., and Walter, H. (2015). GraphVar: a user-friendly toolbox for comprehensive graph analyses of functional brain connectivity. *Journal of Neuroscience Methods* 245, 107–115.
47. Newman, M.E. (2003). The structure and function of complex networks. *SIAM Review* 45, 167–256.
48. Blondel, V.D., Guillaume, J.-L., Lambiotte, R., and Lefebvre, E. (2008). Fast unfolding of communities in large networks. *Journal of Statistical Mechanics: Theory and Experiment* 2008, P10008.
49. Ruhnau, B. (2000). Eigenvector-centrality—a node-centrality? *Social Networks* 22, 357–365.
50. Chung, S., and Abbott, L. (2021). Neural population geometry: an approach for understanding biological and artificial neural networks. *Current Opinion in Neurobiology* 70, 137–144.
51. Beggs, J.M., and Timme, N. (2012). Being critical of criticality in the brain. *Frontiers in Physiology* 3, 163.
52. Friedman, N., Ito, S., Brinkman, B.A., Shimono, M., DeVille, R.L., Dahmen, K.A., Beggs, J.M., and Butler, T.C. (2012). Universal critical dynamics in high resolution neuronal avalanche data. *Physical Review Letters* 108, 208102.
53. Poil, S.-S., Hardstone, R., Mansvelder, H.D., and Linkenkaer-Hansen, K. (2012). Critical-state dynamics of avalanches and oscillations jointly emerge from balanced excitation/inhibition in neuronal networks. *Journal of Neuroscience* 32, 9817–9823.

54. Marshall, N., Timme, N.M., Bennett, N., Ripp, M., Lautzenhiser, E., and Beggs, J.M. (2016). Analysis of power laws, shape collapses, and neural complexity: new techniques and MATLAB support via the NCC toolbox. *Frontiers in Physiology* 7, 250.

55. Ma, Z., Turrigiano, G.G., Wessel, R., and Hengen, K.B. (2019). Cortical circuit dynamics are homeostatically tuned to criticality in vivo. *Neuron* 104, 655–664. e654.

56. Wilting, J., and Priesemann, V. (2018). Inferring collective dynamical states from widely unobserved systems. *Nature Communications* 9, 2325.

Real-Time Multimodal Sensory Detection Using Widefield Hippocampal Calcium Imaging

5

5.1 INTRODUCTION

In Chapter 4, we have seen that the hippocampus encodes information from multiple sensory modalities. The study in this chapter is a further exploration of the previous one, involving the construction of a real-time OBCI to decode spatial, visual, and auditory information from hippocampal signals.

The hippocampus, embedded deep within the temporal cortices of humans and animals, connects with many other brain structures directly or indirectly and plays an important role in memory and cognition. It has been demonstrated to support a spatial cognitive map, providing an environment-centric spatial memory system[1,2]. Apart from encoding spatial information, previous studies have revealed that the hippocampal neuronal network also encodes non-spatial information such as visual[3,4], auditory[5,6,7], olfactory[8,9], gustatory[10], and tactile[11,12]. These observations indicate a more abstract and comprehensive hippocampal cognitive map, formally generating a high-dimensional space for both spatial and non-spatial information. Accurately decoding this mapping in real-time would facilitate direct neural communication in both experimental and potentially clinical scenarios, providing an effective communication channel as a brain–computer interface. However, these non-spatial responses have been found to be spatial context-dependent or task-dependent in almost all previous studies (see O'Keefe and Krupic, 2021 for review)[13]. As a result, it is plausible that such responses may actually reflect the specific location or context in which a particular feature is present rather than the feature itself. It remains unclear whether non-spatial responses can be consistently decoded without concurrent spatial input, especially in real-time scenarios.

To date, only hippocampal spatial information has been resolved in real-time using electrophysiological signals[14,15,16,17] or calcium signals[18]. Detecting hippocampal non-spatial information in real-time is complex and challenging: several recording channels are needed, the sensitivity of hippocampal neurons to non-spatial information is low, analysis pipelines are time-consuming, electrodes often shift, and signals tend to attenuate over time.

Here, we designed an OBCI (Figure 5.1a) based on a single-photon imaging technique (miniscope)[19] in animals traversing a linear track and exposed to light stimuli or tones centred at three different frequencies. While both Tu et al. [18] and our study used different decoders for position reconstruction, a noteworthy innovation in our work is the integration of a Kalman filter after neural network decoding to reduce inherent decoding noise. Importantly, experiments involving non-spatial stimuli were conducted within a small recording chamber to eliminate spatial inputs. Using raw neuronal calcium activity instead of deconvoluted calcium events, and machine learning models, we were able to reconstruct the animal's running trajectory and identify visual and auditory stimuli in real-time (Figure 5.1b–d). We achieved low decoding errors with 9 cm/frame in position reconstruction, 3% in visual stimuli identification, and 17% in auditory stimuli identification. This study presents a demonstration of hippocampal multisensory modality decoding using a multimodal real-time OBCI.

5.2 RESULTS

In our experiments, we utilized a frame rate of 30 Hz, resulting in a frame interval of approximately 33 ms. To ensure consistent frame-by-frame decoding, it was crucial to maintain data processing times below 33 ms to prevent camera data from overflowing the PC buffer. The footprints of neurons are detected offline, and the extraction of raw calcium signals only requires simple

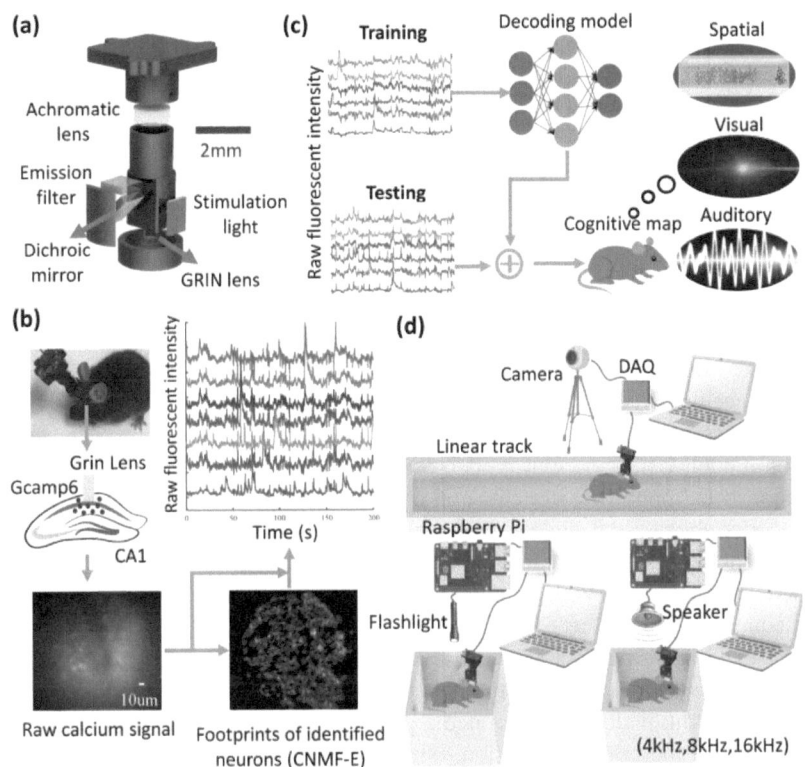

FIGURE 5.1 *In vivo* calcium signal recording and analysis pipelines. (a) Mechanical and optical assembly of the miniscope. (b) A miniscope was used to image the activity of Gcamp6 labelled hippocampal neurons through a GRIN lens. Using the CNMF-E algorithm to identify the spatial footprints of neuronal populations, we extracted the raw calcium activity of detected neurons. (c) A miniscope and a data acquisition board (DAQ) were used to record the hippocampal activity in a mouse. In the position reconstruction experiment, a camera was used to track the location of a mouse traversing a 1.6 m linear track. In the visual/auditory identification experiment, a Raspberry Pi was used to turn a flashlight/speaker on and off to provide visual/auditory stimuli (sinusoidal tones at 4, 8, or 16 kHz) for a mouse within a small recording chamber. (d) Hippocampal activity decoding pipelines. In the training session, the raw fluorescent intensity of sensitive neurons was used to construct the decoding model. This model was then deployed in the real-time session to decode the spatial, visual, and auditory information.

matrix operation. The total processing time, which included image registration and calcium signal extraction, was approximately 2.2 ms. This efficient processing time rendered our system suitable for real-time decoding.

5.2.1 Position Reconstruction

Each mouse ($n = 3$) first underwent a training session in a linear track to construct a position reconstruction model. During the training sessions, we identified a large population of hippocampal neurons in each mouse ($n = 781$, 478, 622). We measured the spatial information content of each neuron and detected many position-sensitive neurons ($n = 322$, 320, 250). An example of the place field map of sensitive neurons is shown in Figure 5.2a. The raw fluorescent intensities of these neurons were used to train the position reconstruction model. Using the four models with a Kalman filter, we could accurately reconstruct running trajectories of the mice. The Gaussian naïve Bayes (GNB) decoder showed the highest decoding error (mean: 22.79 ± 3.42 cm/frame; median:16.00 ± 2.64 cm/frame), while the other three decoders achieved better performance with comparable decoding errors (mean: ~14 cm/frame; median: ~11 cm/frame; Table 5.1).

In the real-time session, the raw fluorescent intensities of location-sensitive neurons detected in the training session were extracted and fed into the decoding model. Additionally, we generated a place field map of the location-sensitive neurons in the real-time session, which revealed distinct and specific firing locations for all neurons, as demonstrated in Figure 5.2b. Our decoding model produced an impressive decoding accuracy, with a low average decoding error of 13.65 ± 0.50 cm/frame (13.20, 13.09, and 14.65 cm/frame, respectively) and a low median error of 9.33 ± 0.67cm/frame (8.00, 10.00, and 10.00 cm/frame). An example of the trajectory reconstruction is shown in Figure 5.2c.

We then examined the noise levels of signals in both the training session and real-time session, but no significant differences were detected (paired Student's t-test did not show significant differences: $t = 0.1362$, df = 79, p-value = 0.8920). Moreover, the cross-correlation of place field maps between the two sessions indicated a peak correlation value at the origin of the coordinate (Figure 5.3a), validating the stability of neuronal firing patterns across sessions. We further evaluated the impact of long short-term memory (LSTM) window sizes on the decoding accuracy and discovered that a 5-frame window size (~0.17 s) provided accurate decoding, whereas a 20-frame window size resulted in substantially poorer performance (Figure 5.3b). We also analysed the effect of neuron numbers on decoding accuracy in the presence of between 20% and 100% of sensitive neurons. We observed that using more neurons in the position reconstruction decreased decoding error (Figure 5.3c) by 15.68%

TABLE 5.1 The mean and median decoding error (cm/frame) in the training session of the position reconstruction experiment using Gaussian naïve Bayes (GNB), support vector machine (SVM), multilayer perceptron (MLP), and long short-term memory (LSTM) decoders

DECODER / MEAN				
/ MEDIAN	MOUSE_1	MOUSE_2	MOUSE_3	ALL (CM/FRAME)
GNB	17.76	20.00	30.60	22.79 ± 3.42
	12	14	22	16.00 ± 2.64
svm	10.45	16.42	13.06	13.31 ± 1.49
	8	12	10	10.00 ± 1.00
mlp	10.29	16.24	16.98	14.50 ± 1.83
	8	12	14	11.33 ± 1.52
lstm	11.16	17.13	16.44	14.91 ± 1.63
	10	12	14	12.00 ± 1.00

Notes: The first row in each cell represents the mean error and the second row represents the median error. The overall error is expressed as mean/median ± standard error mean. Significant differences in average decoding error were observed among different decoders (One-way analysis of variance, $F_{3,6}$ = 5.674, p-value = 0.0347). *Post hoc* tests using Fisher's least significant difference revealed that the average decoding error using the GNB decoder was significantly higher than that using the SVM, MLP, and LSTM decoders, with a p-value of 0.0102, 0.0180, and 0.0220, respectively. No significant differences were found between the other groups: SVM–MLP (p-value = 0.6586), SVM–LSTM (p-value = 0.5563), and MLP–LSTM (p-value = 0.8794). Significant differences in median decoding errors were detected among different decoders (One-way analysis of variance, $F_{3,6}$ = 4.000, p-value = 0.0701). *Post hoc* analysis using Fisher's least significant difference test showed that the median decoding error using the GNB decoder was significantly higher than that using SVM (p-value = 0.0167) and MLP (p-value = 0.0431). No significant differences were found between the other pairs: GNB–LSTM (p-value = 0.0710), SVM–MLP (p-value = 0.4927), SVM–LSTM (p-value = 0.3153), and MLP–LSTM (p-value = 0.7275).

over this range. Finally, we measured the frame processing time for different models and quantities of neurons. With 100% of sensitive neurons employed, the support vector machine (SVM) model was the slowest to process each frame data (~5 ms) compared with the other models that were much faster (~0.2 ms; Figure 5.3d).

5.2.2 Visual Stimuli Identification

We next studied whether the hippocampal ensemble activity could be decoded to identify visual inputs in each frame. To achieve this, we individually tested three mice, each of which underwent a training session to develop the decoding model. In the field of view, we observed 536, 707, and 569 neurons in the three animals, respectively. We measured the information content for each

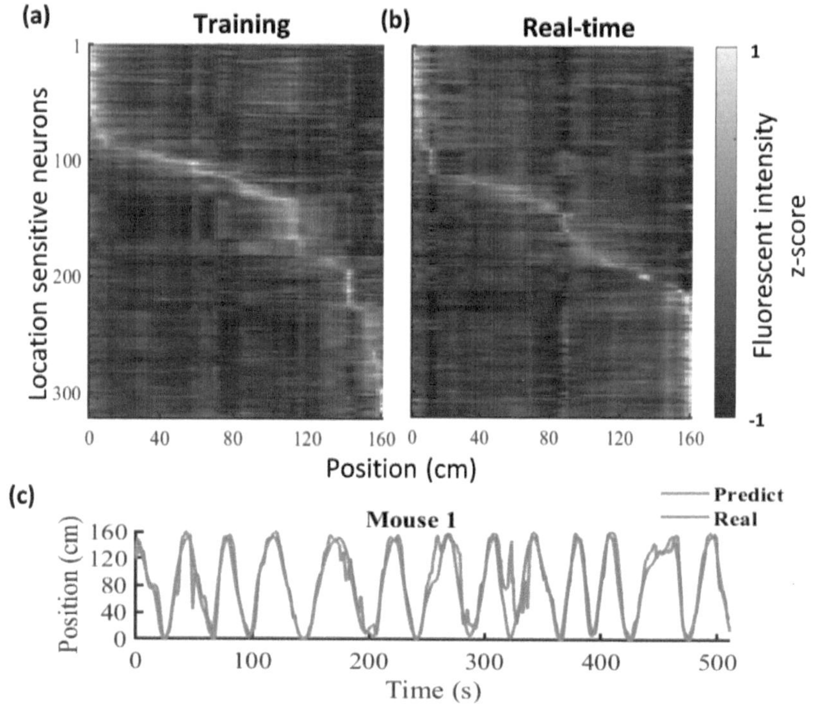

FIGURE 5.2 Firing patterns in position reconstruction experiments. (a) An example of the place field map of sensitive neurons in training sessions. The colour represents the fluorescent intensity z-score. (b) The place field map of sensitive neurons in the real-time session. (c) An example of the mouse's running trajectory reconstruction in a real-time session.

neuron, and detected 204, 392, and 407 sensitive neurons, respectively. We next tested the performance of a GNB decoder, an SVM decoder, and a multi-layer perceptron (MLP) neural network to identify the visual inputs based on the training data. Table 5.2 summarizes the decoding accuracy of each model. In comparison to the GNB decoder (mean error: 20.54% ± 8.70%), the SVM decoder and MLP model both showed very high decoding accuracies (mean error: ~3%). We selected the model with the best performance to apply to each mouse in the subsequent real-time session.

In the real-time session, we used the raw calcium signals of sensitive neurons to predict the visual inputs. Notably, decoding errors were remarkably low in all three mice, with an overall average of 3.36% ± 1.47% (5.47%, 4.07%, and 0.53% in the respective mice). Additionally, the average neuronal activity during light and dark epochs revealed distinct temporal firing patterns (Figure 5.4a–b).

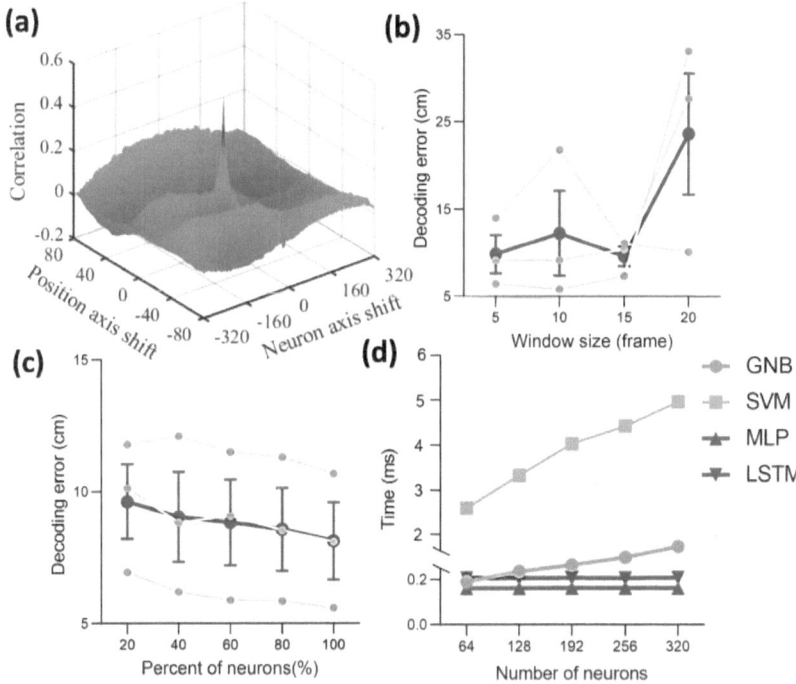

FIGURE 5.3 Position reconstruction decoding models. (a) Cross-correlation of place field maps between the training session and real-time session. (b) Position reconstruction error as a function of the time window size using the long short-term memory (LSTM) model. Blue dots represent the decoding error measured in three mice, and the red dots represent the mean error (Pearson correlation, R^2 = 0.5646, p-value = 0.2486). (c) Position reconstruction error as a function of the percentage of sampled neurons using a multilayer perceptron (MLP) model. Error decreased by incorporating more neurons (Pearson correlation, R^2 = 0.9716, p-value = 0.002). (d) Processing time as a function of the number of sampled neurons in a mouse using different models (Pearson correlation, GNB: R^2 = 0.9926, p-value = 0.0003; SVM: R^2 = 0.9876, p-value = 0.0006; MLP: R^2 = 0.7701, p-value = 0.0505; long LSTM: R^2 = 0.5103, p-value = 0.1752).

The noise levels of the signals in the training sessions and the real-time sessions did not show significant differences (paired Student's t-test did not show significant differences: t = 0.1362, df = 79, p-value = 0.8920). Intriguingly, the maximum cross-correlation value of firing rate maps between two sessions was not at the origin of the coordinate axes, indicating that neuronal temporal firing patterns were not highly consistent across sessions (Figure 5.5a). This possibly implies that visual stimuli not only have temporal

TABLE 5.2 The average decoding error ratio in the training session of the light stimuli experiment using GNB, SVM, and MLP decoders

DECODER	MOUSE_1	MOUSE_2	MOUSE_3	ALL
GNB	22.95%	34.28%	4.40%	20.54% ± 8.70%
svm	3.99%	4.01%	1.22%	3.08% ± 0.92%
mlp	5.36%	4.63%	0.73%	3.57% ± 1.43%

Notes: The overall error ratio is expressed as mean ± standard error mean. Significant differences in average decoding error were observed among different decoders (One-way analysis of variance, $F_{2,4} = 17.39$, p-value = 0.0106). *Post hoc* analysis using Fisher's *least significant difference* test revealed that the average decoding error with the GNB decoder was significantly higher than that with SVM (p-value = 0.0067) and MLP (p-value = 0.0073). However, no significant differences were found between SVM and MLP decoders (p-value = 0.9071).

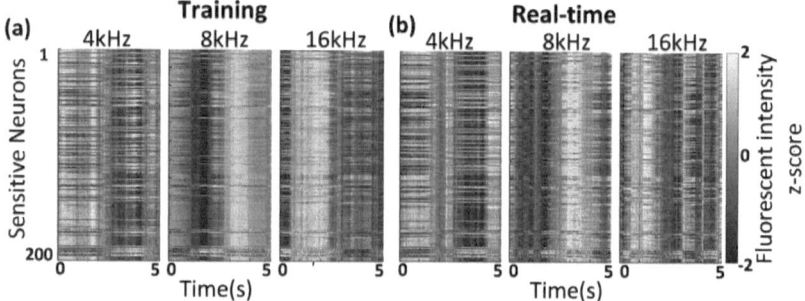

FIGURE 5.4 Firing patterns in visual stimuli experiments. (a) The visual stimuli evoked neuronal activity of sensitive neurons in light and dark epochs in the training session. The colour represents the fluorescent intensity *z*-score. The flashlight was turned on or off at time "0". (b) The visual stimuli evoked neuronal activity of sensitive neurons in light and dark epochs in the real-time session.

effects on neuronal activity, but also affect the functional connectivity of neuronal populations. On the other hand, this could be attributed to sensory adaptation. The cross-correlation measured the similarity of the neuronal firing rate maps between the training session and the real-time session. It is important to note that this method primarily captures the fundamental characteristics of population activity and cannot discern deep features within the neuronal firing pattern. Considering that the neural network decoding model was able to provide accurate decoding even in this scenario, it may imply that neuronal activity related to non-spatial stimuli is better characterized by deeper features. This could potentially explain why some previous studies have reported a low sensitivity of hippocampal neurons to these non-spatial stimuli[13], possibly due to the analysis method. Increasing the number of neurons used in the decoding

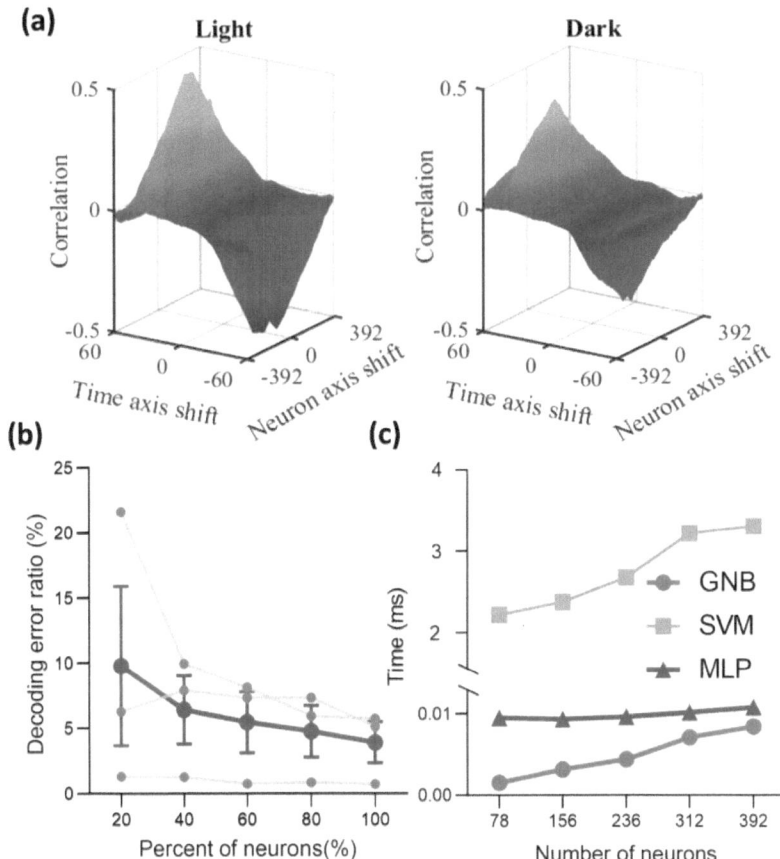

FIGURE 5.5 Visual stimuli decoding models. (a) Cross-correlation of firing rate maps between the training session and real-time session in light and dark environments. The unit of time axis shift is the frame, which is around 33 ms. (b) Decoding error as a function of the percentage of sampled neurons using a MLP model. Blue dots represent the decoding error measured on three mice, and the red dots represent the mean error. Error decreased considering more neurons (Pearson correlation, $R^2 = 0.8702$, p-value = 0.0207). (c) Processing time as a function of the number of sampled neurons in a mouse using different models (Pearson correlation, GNB: $R^2 = 0.9853$, p-value = 0.0008; SVM: $R^2 = 0.9523$, p-value = 0.0045; MLP: $R^2 = 0.8162$, p-value = 0.0355).

process resulted in higher decoding accuracy, as observed in position recon-struction experiments (Figure 5.5b). As the decoding model was simpler than that used in position reconstruction experiments, the MLP model only required about 0.01 ms to process the data in each frame (Figure 5.5c).

5.2.3 Auditory Stimuli Identification

Three mice were exposed to pure sinusoidal tones at frequencies of 4, 8, and 16 kHz to investigate if the activity of hippocampal neurons could be decoded to differentiate between the auditory stimuli frequencies. We observed 534, 619, and 599 neurons in each mouse. Initially, we used a GNB decoder, an SVM decoder, and an MLP neural network model to identify the auditory input in each frame during the training session. However, none of these models were able to make accurate predictions. Table 5.3 summarizes the decoding error of each model. We next sectioned the data into epochs according to the fre-quency of stimuli, where each epoch contained a 2 s sound-on period followed by a 3 s sound-off period. We detected 198, 227, and 230 sensitive neurons in each mouse. An example of the average neuronal activity of sensitive neurons is shown in Figure 5.6a. GNB, SVM, and MLP models are less effective in handling temporal multidimensional data. A common approach to analyse such data is to flatten or reshape the inputs. However, this method significantly increases dimensionality, which brings about challenges such as the curse of dimensionality, heightened computational complexity, and the potential loss of spatial information in the features. Given these challenges, along with

TABLE 5.3   The average decoding error ratio in the training session of the sound stimuli experiment using GNB, SVM, MLP, and convolutional neural network (CNN) decoders

DECODER	MOUSE_1	MOUSE_2	MOUSE_3	ALL
GNB	71.09%	77.50%	81.34%	76.64% ± 2.59%
svm	42.81%	39.85%	40.05%	40.91% ± 0.82%
mlp	39.33%	37.16%	37.84%	38.11% ± 0.55%
CNN	17.67%	22.76%	22.89%	21.11% ± 1.48%

Notes: The overall error ratio is expressed as mean ± standard error mean. Significant differences in average decoding error were observed among different decoders (One-way analysis of variance, $F_{3,6}$ = 154.5, p-value < 0.0001). Fisher's least significant difference $post$ hoc test revealed that the average decoding error using the GNB decoder was significantly higher than that using SVM, MLP, and CNN decoders (p-value < 0.0001). However, no significant differences were found between SVM and MLP decoders (p-value = 0.3333). Additionally, decoding accuracy using CNN decoder was significantly higher than that using SVM and MLP decoders, with a p-value of 0.0003 and 0.0007, respectively.

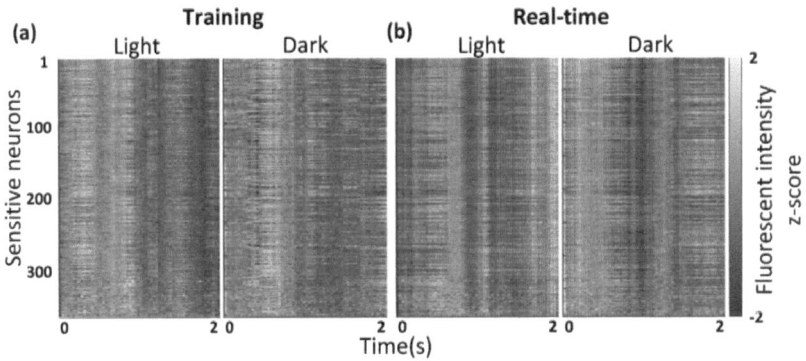

FIGURE 5.6 Firing patterns in auditory stimuli experiments. The mice were exposed to sound stimuli at three different frequencies. (a) The sound stimuli evoked neuronal activity of sensitive neurons in 4, 8, and 16 kHz epochs during the training session. The colour represents the fluorescent intensity z-score. The speaker was turned on at the time "0" and turned off at the time "2". (c) An example of the CNN decoding performance in the training session. The data were sectioned into 5-s time–length epochs according to the frequency of the stimuli. (b) The sound stimuli evoked neuronal activity of sensitive neurons in the real-time session.

substantial optimization time consumption, we did not use these models for further analysis. We then trained a convolutional neural network (CNN) model to identify the frequency of each epoch and achieved high decoding accuracy with an error rate of 17.67%, 22.76%, and 22.89% in each mouse (mean ± SEM: 21.11% ± 1.48%).

In the real-time session, the CNN model constructed in the training session was deployed to decode the raw calcium activity of sound sensitive neurons. The resulting average temporal firing patterns exhibited similarity to those observed in the training session (Figure 5.6b). For each mouse, the decoding error ratio was 13.25%, 20.78%, and 19.48% respectively (mean ± SEM: 17.83% ± 2.32%).

In the study, noise levels in the signals of both the training and real-time sessions were found to be similar (paired Student's t-test did not show significant differences: $t = 1.226$, df=79, p-value = 0.2238). Cross-correlation analysis of firing rate maps between two sessions revealed multiple peaks, with a dominant peak located at the origin, suggesting relatively consistent temporal firing patterns across sessions (Figure 5.7a). Likewise, the highest decoding accuracy was achieved using all sensitive neurons (Figure 5.7b) and the model consumed a short period of ~1.6 ms to process the 5 s epoch data using the CNN model (Figure 5.7c).

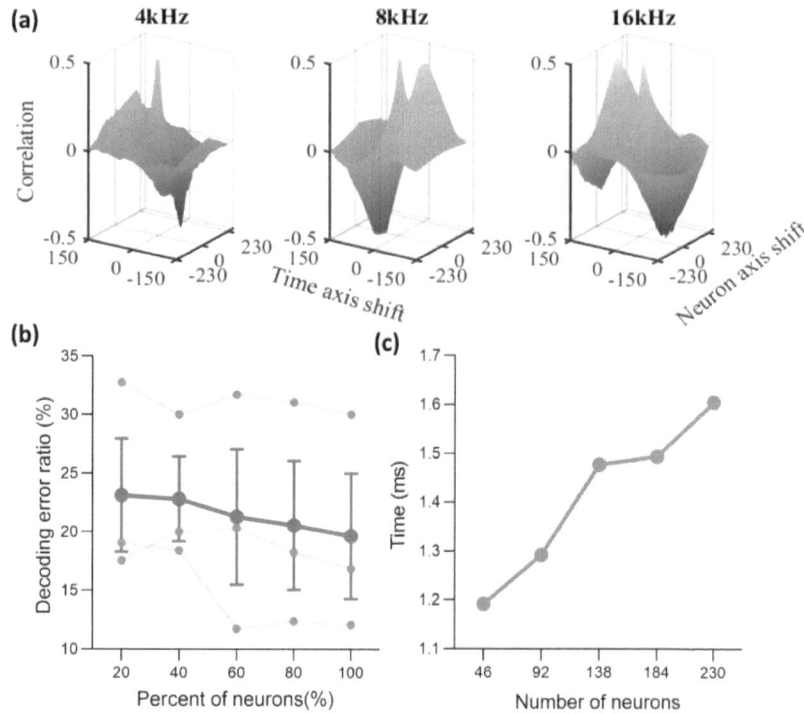

FIGURE 5.7 Auditory stimuli decoding models. (a) Cross-correlation of firing rate maps between the training session and real-time session in three environments. The unit of time axis shift is the frame, which is around 33 ms. (b) Decoding error as a function of the percentage of sampled neurons using the CNN model. Blue dots represent the decoding error measured on three mice, and the red dots represent the mean error. Error decreased considering more neurons (Pearson correlation, $R^2 = 0.9717$, p-value = 0.002). (c) Processing time as a function of the number of sampled neurons in a mouse using the CNN model (Pearson correlation, GNB: $R^2 = 0.9518$, p-value = 0.0046).

5.3 DISCUSSION

We have decoded multiple sensory modalities from hippocampal neuronal activity in mice, and our proposed method can recognize certain pre-trained patterns that are associated with specific behaviours. Our low-latency end-to-end analysis pipeline provided accurate decoding of both hippocampal spatial and non-spatial information, which allowed a real-time readout of these internal cognitive states, delivering a functional cognitive interface. The hippocampus supports a diverse cognitive map that incorporates both spatial and non-spatial information and therefore represents a useful target for decoding multimodal information, which is particularly suited for multisensory decoding compared with other primary sensory brain regions.

The projection pathways of sensory information to the hippocampus differ anatomically. Spatial information mainly targets the dorsal and posterior hippocampus[20]. Guger et al., Sodkomkham et al., and Hu et al. have reconstructed rat running trajectories in real-time using small sets of hippocampal neurons recorded electrophysiologically[14,15,17]. Non-spatial information has been reported to largely flow into the ventral and anterior hippocampus[20]. Visual signals project to the hippocampus from the visual cortex through a multisynaptic pathway[21,22]. Previous studies found that hippocampal activity could represent visual stimuli[4], and neurons in the hippocampus and visual cortex showed synchrony in visual stimuli experiments[23]. The transmission pathways between the hippocampus and auditory cortex are complex. It is believed that there are two major pathways – the lemniscal pathway and the non-lemniscal pathway, and the auditory signalling to the hippocampus has likely undergone several integrative stages[7,24]. Neurons in the hippocampus of rodents have been reported to show a certain degree of sound sensitivity[6]. All these non-spatial sensitivity investigations made use of low channel counts electrophysiological recording methods but did not show whether the activity could be decoded in real-time. In comparison, our optical BCI enabled recording of larger neuronal populations, and we demonstrated that these non-spatial modalities can be detected in real-time with high fidelity.

5.3.1 Decoding Models

We tested different machine learning models to decode the hippocampal cognitive map including the GNB decoder, SVM decoder, MLP neural network model, LSTM neural network model, and CNN model. In all cases, the GNB decoder showed the highest decoding error, which might be due to the non-linearity of the hippocampal neuronal network and the inherently complex signal composition. Alternatively, the determination of adequate priors may have been suboptimal.

In both the position reconstruction and visual stimuli identification experiments, the MLP neural network model and SVM decoder showed similar decoding performance in all animals. Compared with the MLP neural network model, the SVM decoder has the advantage of optimizing fewer hyperparameters and can be trained in an online mode[25,26], which avoids a separate training session. The LSTM algorithm is an artificial recurrent neural network that can achieve good prediction accuracy from time-series data[27,28]. It simulates a biologically relevant model of neuronal activity processing. However, it did not show the best performance in position reconstruction experiments, which was somewhat surprising. A possible explanation may be the slow kinetics of neuronal calcium activity. Action potentials cause calcium influx and efflux in excitable cell bodies, and the depolarization-evoked neuronal firing has a long-lasting effect on calcium activity. This indicates that the calcium activity in the current frame inherits partial information encoded in previous frames, which might weaken the strength of memory units in the model. It is known that incorporating too much old information in an LSTM neural network model causes a drop in its performance because that information may not be relevant or useful or may introduce unwanted noise[29].

We showed that spatial and visual information could be decoded accurately in each recorded frame of activity. However, a relatively long-time window was needed to decode the auditory information. This may be due to various encoding mechanisms for different types of sensory information. Indeed, auditory-evoked neuronal activation has been reported to exhibit variable latencies[6,30], resulting in long-lasting temporal firing patterns that would be consistent with this hypothesis. Another explanation may be the relatively low absolute sensitivity of CA1 neurons to auditory stimuli[5,6], necessitating the highly optimized analysis and longer recording periods.

5.3.2 Optical Brain–Computer Interfaces (OBCIs)

Conventional intracranial BCIs use electrode arrays to record neuronal action potentials or local field potentials to detect internal states. More recently, the potential for calcium imaging using multiphoton imaging of neuronal ensembles as a BCI has been examined[31,32]. The relationship between neuron action potentials and calcium activity is complex. An action potential activates voltage-gated calcium channels, eliciting a nonlinear rise in intracellular calcium concentration. Although dynamics of calcium activity are relatively slow, it has been shown to track action potential frequency[33]. In our experiments, we used GCaMP6f with relatively fast kinetics (~50 ms temporal resolution)[34]. Thus, the temporal resolution of calcium signals seems functionally comparable to electrical signals.

Clancy et al. utilized volitional control of a small number of neurons identified with two-photon imaging that could indirectly modulate the sound from a speaker[31]. Trautmann et al. implemented a two-photon OBCI in a head-fixed macaque for detection of the animal's arm motion[32]. In contrast, our technique uses single-photon imaging, which yields poorer signal quality compared to two-photon imaging since it also collects neuronal firing activity of out-of-focus cells as well. However, it is important to note that this additional neuronal firing information has the potential to enhance decoding accuracy. Additionally, this workflow is low-cost and does not require bulky equipment. Interestingly, Chen et al. recently implemented an approach utilizing hippocampal calcium signals and simple linear decoders to estimate the positional information of Long-Evans rats within a linear track[35]. Despite the generally higher spatial specificity of rat place cells in comparison to mouse, the researchers reported a decoding error of 30 cm/frame, which is higher than the 9 cm/frame error observed in our experiments. Additionally, the authors implemented a relatively large spatial bin size of 20 cm in their analysis. Tu et al. reconstructed an animal's position in a sensory-cued treadmill using single-photon calcium imaging datasets[18]. Their maximum-likelihood estimation method demonstrated good decoding accuracy, achieving a decoding error of around 6 cm. However, it is not clear whether this measurement refers to 6 cm per second or 6 cm per frame. Comparing decoding accuracy between their study and ours presents challenges due to differences in experimental design. These distinctions encompass varying recording sampling frequencies, which are known to impact calcium signal quality, as well as differences in experimental apparatus. It is important to note that the animal had to maintain a stable running speed on the treadmill, and running speed can impact place cell activity[36]. Furthermore, discrepancies in our training methods further complicate the comparison, as it is difficult to ascertain whether the animals were trained to the same level.

In the experiment, we used a relatively short 1.6m linear track. This limitation is attributed to the confined space within the experimental room, with the issue of cable handling posing a more substantial challenge. Managing longer cables can become challenging as the linear track extends. However, one solution to address the cable-related challenges is to use a wire-free miniscope[37], which has recently become available on the market.

In summary, we have successfully decoded spatial, visual, and auditory information from the mouse hippocampus using widefield single photon imaging in real-time, constituting an effective OBCI. The end-to-end OBCI system proposed here presents a proof of concept for decoding the hippocampal cognitive map in real-time. It expands the method and opportunity to study the activity of hippocampal neuronal ensembles and will be helpful for future content-specific closed-loop BCI experiments. Furthermore, it provides an approach for multisensory modality decoding, which may be applied in clinical applications and scientific research in the future. Additionally, future studies exploring the application of OBCI to decode action behaviours hold particular promise, offering potential benefits to patients affected by functional impairments.

5.4 METHODS

5.4.1 Miniscope System Design

Our miniscope optical system was developed based on the previous design from UCLA[19]. The UCLA design is a compact fluorescence microscope consisting of a stimulation light source, half ball lens, excitation filter, dichroic mirror, GRIN lens, emission filter, and achromatic lens. We made changes to this system by replacing the original achromatic lens with a shorter focal length of 7.5 mm (45407, Edmund Optics) and the GRIN lens with a pitch value of 0.23 (64520, Edmund Optics). This modified optical system has a shorter focal length, lower magnification, and larger field of view, which are desirable features for BCI applications (Figure 5.8). A shorter focal length in the optical system can decrease the overall height of the miniscope. Lower magnification can reduce the requirement for utilizing all available CMOS sensor pixels, enabling more efficient real-time processing. A larger field of view can provide access to more neurons, improving decoding accuracy. In our practical implementation, it was observed that the inclusion of an excitation filter and half ball lens was not imperative for capturing high-quality images. The stimulation LED exhibits excellent monochromaticity and collimation, and the dichroic mirror additionally enhances the purity of the stimulation light. Removing these optics further reduces the size and weight of the miniscope. Figure 5.8c depicts a comparison of the imaging quality obtained with and without an excitation filter and a half ball lens.

5.4.2 Animals and Surgery

Two stereotaxic surgeries were conducted on mice under anaesthesia (isoflurane: 3%–5% induction, 1.5% maintenance). To label hippocampal neurons, mice were unilaterally injected with 500 nL of pAAV.Syn.GCaMP6f. WPRE.SV40 virus (#100837-AAV1, AddGene) in the dorsal CA1 (right brain hemisphere, 2.1 mm posterior to the bregma, 2.1 mm lateral to the midline, and 1.65mm ventral from the surface of the skull). One week following the injection, a circular craniotomy, 2 mm in diameter, was performed next to the injection site (2.1 mm posterior to the bregma, 1.6 mm lateral to the midline). The cortex above the corpus callosum was aspirated using a 27-gauge blunt needle attached to a vacuum pump, and a 1.8 mm diameter GRIN lens was implanted at a depth of 1.35 mm from the surface of the skull. The lens was fixed in place with cyanoacrylate glue and dental acrylic and protected with silicone adhesive

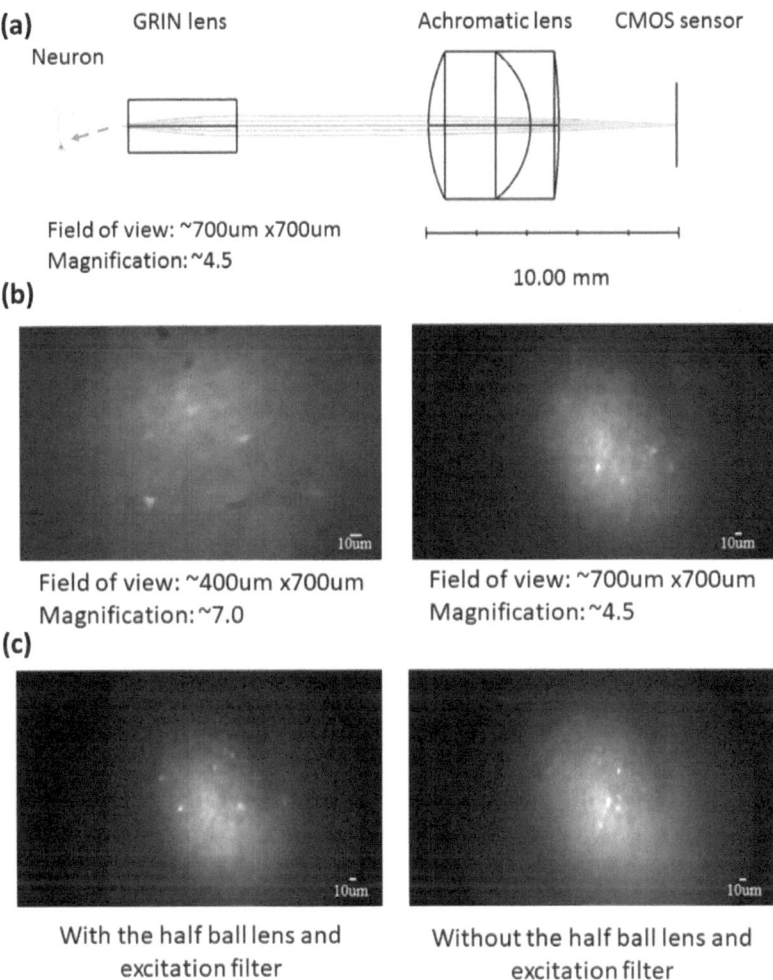

FIGURE 5.8 Zemax Simulation and imaging of hippocampal CA1 neurons in freely behaving mice. (a) Zemax simulation of emission path of our modified optical system. (b) The modified optical system has a smaller magnification in comparison to the conventional one and provides a larger field of view. (c) Imaging with/without a half-ball lens and an excitation filter shows similar imaging quality.

(Dragon Skin® Series). Mice were given analgesics (carprofen: 5 mg/kg; dexamethasone: 0.6 mg/kg) and enrofloxacin water (1:150 dilution, Baytril®) to recover for 7 days. Neuronal calcium activity was measured 4–5 weeks later. After finding the best field of view, a baseplate was cemented on the head and a plastic cap was locked into the baseplate to protect the lens[38].

5.4.3 Information Content Analysis and Neuronal Sensitivity Identification

Information content can be used to quantify the precision of neuronal coding with a large value indicating more precise coding. In the position reconstruction experiment, the definition of the information content is similar to that described previously[39,40], but we changed the measurement of neuronal firing rate to neuronal fluorescent intensity to adapt to our recording technique,

$$ I = \frac{r_i}{\bar{r}} \log_2 \frac{r_i}{\bar{r}}, \bar{r} = \sum_{i=1}^{k} p_i r_i, \tag{5.1} $$

where I represents the information content, i is the spatial bin index (the linear track is divided into several 2 cm bins), k is the number of spatial bins, p_i is the probability of occupancy of the ith bin, r_i is the average fluorescent intensity in the ith bin, and $\bar{r}$ is the overall mean fluorescent intensity.

To identify neurons responsive to place, the animal's location was shuffled, and the information content was recomputed for each shuffle. This step repeated 1000 times, and a neuron was marked as a location sensitive neuron if its information content in the unshuffled trial exceeded 95% of the shuffled trials.

The definition of the information content for visual and auditory stimuli was the same, but the data binning was implemented in the temporal domain. There were two states (light or dark) in the visual experiments and three states denoted by three different frequencies in the auditory experiments. The same shuffling method was used to detect light and sound sensitive neurons.

5.4.4 Calcium Activity Decoder

The performance of several models to decode neural signals was evaluated and compared. These included a GNB decoder, an SVM decoder, an MLP neural network model, a LSTM neural network model, and a CNN model. The decoders were constructed and implemented on the Python platform using the scikit-learn toolkit and TensorFlow library[41,42].

The GNB decoder is a type of probabilistic-based prediction algorithm based on Bayes' theorem. In the position reconstruction experiment, raw fluorescent intensities of location sensitive neurons were first normalized to remove the mean and scaled to unit variance. Given a sequence of fluorescent intensities from location sensitive neurons in each frame, the estimated position $\hat{y}$ was defined as,

$$\hat{y} = \underset{y}{\arg\max} \; P(y)\prod_{i=1}^{N}P(x_i \mid y), P(x_i \mid y) = \frac{1}{\sqrt[2]{2\pi\sigma_y^2}}e^{-\frac{(x_i-\mu_y)^2}{2\sigma_y^2}} \qquad (5.2)$$

where $P(y)$ is the probability of occupancy of bin y, x_i is the normalized fluorescent intensity of the ith neuron, and μ_y and σ_y represent the mean and standard deviation of the normalized fluorescent intensity of the ith neuron at bin y, respectively. In visual stimuli experiments, y represents either light-bin or dark-bin. In auditory stimuli experiments, y represents the bins with different frequencies.

The SVM decoder performs classifications by constructing a set of hyperplanes that maximizes the margins between different classes. The SVM model was trained and constructed using the normalized input data with a nonlinear radial basis function kernel. The kernel width, gamma, was set to be the reciprocal of the number of input features. A cost parameter "C" was optimized by applying a grid search technique with fivefold cross-validation.

An MLP is an artificial neural network that is commonly used for solving prediction and classification problems especially when the input data is not linearly separable. We built an MLP model with one input layer receiving normalized data, two hidden layers activated by a rectified linear unit (ReLU) function, and one output layer with softmax activation. An Adam optimizer and a categorical cross-entropy loss function were used to compile the model. The batch size was set to 32 and all the other hyperparameters, including the number of nodes in hidden layers, learning rate, and the number of epochs, were optimized by implementing a grid search method using fivefold cross-validation.

An LSTM model is a type of recurrent neural network that can use internal memory to process sequences of data with variable lengths and is extremely useful for time series forecasting. The LSTM model was implemented to reconstruct the animal's running trajectory in our experiments. The model consisted of one input layer, two fully connected LSTM layers activated by a ReLU function, a dropout layer (dropout rate: 0.2) that prevented overfitting,

and one output layer with a softmax activation function. In position reconstruction experiments, the output of the model was marked as the animal's
current location bin, and different lengths of normalized time-series data were
evaluated to build and train the model. The same method as the MLP model
was used to compile and optimize the model.

A CNN is an artificial neural network that has been frequently implemented
in image processing, but also shows good performance for time-series data.
It is designed to detect spatial hierarchies of features in the input data. The
CNN model was tested to decode the hippocampal activity in auditory stimuli
experiments. The inputs were the time series of raw fluorescent intensities
from sensitive neurons. The model contained two convolutional layers using
ReLU activations, with a max-pooling layer added after each convolutional
layer for dimensionality reduction. A dropout layer (dropout rate: 0.2) was
then concatenated to prevent overfitting. Finally, a fully connected layer with
a softmax activation function was added to output the probability distribution for each class. A grid search method was performed to determine the
hyperparameters that yielded the highest decoding accuracy.

In our decoding methods, such as the GNB decoder, SVM decoder, and
MLP neural network decoder, we used single-frame data as the input. However,
when employing the LSTM neural network model, we tested different window
lengths and determined that a temporal window of five frames typically
provided accurate decoding. As the auditory stimuli have a long temporal
effect on neuronal activity, a temporal window of 150 frames is used as inputs
when using the CNN model.

5.4.5 Kalman Filter

In the position reconstruction experiment, a Kalman filter was used to reduce
the decoding noise in the outputs of the decoder. The Kalman filter is one of
the most widely used methods for position tracking. To estimate the state $\hat{x}_t$
[position and velocity] of the animal at time t, the estimation processes are
defined as:

Time update:

$$\hat{x}_{\bar{t}} = \mathbf{A}\hat{x}_{t-1} + \mathbf{B}u_{t-1}, \; P_{\bar{t}} = \mathbf{A}P_{t-1}\mathbf{A}^T + \mathbf{Q} \tag{5.3}$$

Measurement update:

$$K_t = P_{\bar{t}}\mathbf{H}^T\left(\mathbf{H}P_{\bar{t}}\mathbf{H}^T + \mathbf{R}\right)^{-1}, \quad P_t = \left(\mathbf{I} - K_t\mathbf{H}\right)P_{\bar{t}}, \; \hat{x}_t = \hat{x}_{\bar{t}} + K_t\left(z_t - \mathbf{H}\hat{x}_{\bar{t}}\right) \tag{5.4}$$

$$\mathbf{A} = \begin{bmatrix} 1 & \Delta t \\ 0 & 1 \end{bmatrix}, \mathbf{B} = \begin{bmatrix} \frac{1}{2}\Delta t^2 \\ \Delta t \end{bmatrix}, \mathbf{H} = \begin{bmatrix} 1 & 0 \end{bmatrix} \tag{5.5}$$

where $\hat{x}_{\bar{t}}$ is the prior estimation of the state, $P_{\bar{t}}$ is the prior transition covariance, K_t is the Kalman gain, P_t is the updated transition covariance, $\hat{x}_t$ is the updated state, u_{t-1} is the acceleration, $\mathbf{Q}$ is the transition covariance, and $\mathbf{R}$ is the observation covariance. The sampling frequency is 30 fps, so Δt equals 1/30. The initial values of the state were set to be zero and the transition covariance matrix was set to be identity. The values of $\mathbf{Q}$ and $\mathbf{R}$ were set to be 0.0001 and 1, respectively.

5.4.6 Experimental Procedures and Analysis Method

All experiments consisted of two sessions: (1) a training session and (2) a real-time session. The training session was implemented offline to construct the decoding model, which was then used to decode neuronal activity in a real-time session.

5.4.7 Real-Time Position Reconstruction Experiment

Prior to the commencement of this experiment, mice were kept on dietary restriction and maintained at 85% of free-feeding body weights. The mice were trained to traverse a 1.6-m linear track for food rewards with miniscopes attached and were required to complete 12 trials of traversing each day for 1 week.

On the day of the training session, hippocampal neuronal activity was recorded when the mouse traversed a linear track for 12 trials. The sampling frequency of the miniscope was set to 30 fps and animal's position was tracked using a video camera that was synchronized with the recording system. After recording, the translational frame shifting was corrected using a cross-correlation-based image registration algorithm[43]. The spatial footprints of neurons in the field of view were detected by implementing a CNMF-E algorithm[44] with the centroid of each neuron and a small surrounding area used to measure activity. For each neuron, the raw calcium activity was defined as the average fluorescent intensity of the centroid pixel and the surrounding eight pixels. We then detected location-sensitive neurons using a shuffling method (see Methods). Next, signals from location-sensitive neurons were used to construct a decoding model to optimize position identification. The performance of the four decoders described above (GNB, SVM, MLP, and LSTM) was evaluated and compared. Finally, the decoding noise from the outputs of the decoder was reduced with a Kalman filter. The best decoder, together with the Kalman filter, was subsequently deployed in the real-time sessions.

In the real-time session, the mouse traversed the linear track for 12 trials. The position of the mouse was tracked using a video camera to assess the decoding accuracy. The same image registration method was implemented to align the image, and the raw calcium activity of position sensitive neurons detected in the training session was extracted and fed into the decoder to reconstruct the animal's running trajectory in real-time.

5.4.8 Visual and Auditory Stimuli Identification

Hippocampal activity evoked by visual stimuli and auditory stimuli was studied separately. The mouse was put in a small opaque recording chamber and was exposed to either a flashlight or a speaker mounted above the chamber.

On the day of the training session, the mouse was placed in the chamber 15 min before recording to acclimate. A Raspberry Pi 3 (Model B) board controlled a flashlight or a speaker. In the visual stimuli experiment, the flashlight was turned on for 2 s and off for 2 s alternately 150 times. In the auditory stimuli experiment, the speaker played three different frequency tones (4, 8, and 16 kHz) randomly, 225 times with an activation-period of 2 s and a mute-period of 3 s. The data were analysed using the same procedure described in the position reconstruction experiment: (1) image registration, (2) neuron centroid detection, (3) raw calcium activity extraction, (4) sensitive neuron detection, and (5) decoder construction. The performance of the GNB decoder, SVM decoder, and MLP neural network model were assessed and compared. In the auditory stimuli experiment, none of these three decoders showed outstanding performance to identify the stimulus in each frame (see Results). A possible reason was that auditory information was encoded in temporal sequences, so we divided the data into several epochs for further analysis. The time length of the epoch was 5 s, which included the speaker activation and deactivation periods. Finally, a CNN model was constructed to decode the epoch data.

In the real-time session, the mouse was exposed to 150 light stimuli or 80 sound stimuli. The raw calcium activity of stimuli sensitive neurons detected in the training session was extracted after image registration and provided to the decoder. The MLP model or SVM model that showed the best performance in the training session was deployed in visual stimuli experiments and a CNN model was deployed in auditory stimuli experiments to do the subsequent real-time decoding.

5.4.9 Noise Level and Firing Rate Map Similarity Analysis

The noise level was used to characterize the noise coupled to the calcium dynamics. Raw calcium activity was first processed with a zero-phase infinite impulse response lowpass filter (1 Hz cut-off frequency, filter order: 20), and the signal noise level was defined as the difference between the raw calcium activity and the filtered activity.

To compare the consistency or the similarity of the firing rate map between the training session and the real-time session, normalized cross-correlation was measured[45]

$$R(u,v) = \frac{\sum_{xy} \left[m(x,y) - \bar{m}_{u,v} \right] \left[n(x-u, y-v) - \bar{n} \right]}{\sqrt{\sum_{xy} \left[m(x,y) - \bar{m}_{u,v} \right]^2 \sum_{xy} \left[n(x-u, y-v) - \bar{n} \right]^2}} \tag{5.6}$$

where m and n represent the firing rate maps in the training session and the real-time session, x and y represent the pixels in the maps, and u and v represent the pixel shift along different dimensions in each map, respectively.

5.4.10 Statistics and Reproducibility

The statistical analysis was conducted utilizing Graphpad PRISM version 7 software. Statistical significance was determined at p-value < 0.05. Statistical significance was determined through a two-tailed paired Student's t-test or one-way ANOVA. The proposed analysis procedure was tested on three animals. Due to experimental design, the experimenter was not able to remain blinded to the manipulations carried out during the experiment.

REFERENCES

1. O'keefe, J. and Nadel, L., 1978. *The Hippocampus as a Cognitive Map.* Oxford: Oxford University Press.
2. Spiers, H.J., 2020. The hippocampal cognitive map: one space or many?. *Trends in Cognitive Sciences*, 24(3), pp.168–170.

3. Acharya, L., Aghajan, Z.M., Vuong, C., Moore, J.J. and Mehta, M.R., 2016. Causal influence of visual cues on hippocampal directional selectivity. *Cell*, *164*(1–2), pp.197–207.

4. Liu, Y.Z., Wang, Y., Tang, W., Zhu, J.Y. and Wang, Z., 2018. NMDA receptor-gated visual responses in hippocampal CA1 neurons. *The Journal of Physiology*, *596*(10), pp.1965–1979.

5. Moita, M.A., Rosis, S., Zhou, Y., LeDoux, J.E. and Blair, H.T., 2003. Hippocampal place cells acquire location-specific responses to the conditioned stimulus during auditory fear conditioning. *Neuron*, *37*(3), pp.485–497.

6. Itskov, P.M., Vinnik, E., Honey, C., Schnupp, J. and Diamond, M.E., 2012. Sound sensitivity of neurons in rat hippocampus during performance of a sound-guided task. *Journal of Neurophysiology*, *107*(7), pp.1822–1834.

7. Xiao, C., Liu, Y., Xu, J., Gan, X. and Xiao, Z., 2018. Septal and hippocampal neurons contribute to auditory relay and fear conditioning. *Frontiers in Cellular Neuroscience*, *12*, p.102.

8. Komorowski, R.W., Manns, J.R. and Eichenbaum, H., 2009. Robust conjunctive item–place coding by hippocampal neurons parallels learning what happens where. *Journal of Neuroscience*, *29*(31), pp.9918–9929.

9. Taxidis, J., Pnevmatikakis, E.A., Dorian, C.C., Mylavarapu, A.L., Arora, J.S., Samadian, K.D., Hoffberg, E.A. and Golshani, P., 2020. Differential emergence and stability of sensory and temporal representations in context-specific hippocampal sequences. *Neuron*, *108*(5), pp.984–998.

10. Ho, A.S., Hori, E., Thi Nguyen, P.H., Urakawa, S., Kondoh, T., Torii, K., Ono, T. and Nishijo, H., 2011. Hippocampal neuronal responses during signaled licking of gustatory stimuli in different contexts. *Hippocampus*, *21*(5), pp.502–519.

11. Pereira, A., Ribeiro, S., Wiest, M., Moore, L.C., Pantoja, J., Lin, S.C. and Nicolelis, M.A., 2007. Processing of tactile information by the hippocampus. *Proceedings of the National Academy of Sciences*, *104*(46), pp.18286–18291.

12. Gener, T., Perez-Mendez, L. and Sanchez-Vives, M.V., 2013. Tactile modulation of hippocampal place fields. *Hippocampus*, *23*(12), pp.1453–1462.

13. O'Keefe, J. and Krupic, J., 2021. Do hippocampal pyramidal cells respond to nonspatial stimuli?. *Physiological Reviews*, *101*(3), pp.1427–1456.

14. Guger, C., Gener, T., Pennartz, C., Brotons-Mas, J., Edlinger, G., Badia, B.I., Schaffelhofer, S., Verschure, P. and Sanchez-Vives, M.V., 2011. Real-time position reconstruction with hippocampal place cells. *Frontiers in Neuroscience*, *5*, p.85.

15. Sodkomkham, D., Ciliberti, D., Wilson, M.A., Fukui, K.I., Moriyama, K., Numao, M. and Kloosterman, F., 2016. Kernel density compression for real-time Bayesian encoding/decoding of unsorted hippocampal spikes. *Knowledge-Based Systems*, *94*, pp.1–12.

16. Ciliberti, D., Michon, F. and Kloosterman, F., 2018. Real-time classification of experience-related ensemble spiking patterns for closed-loop applications. *Elife*, *7*, p.e36275.

17. Hu, S., Ciliberti, D., Grosmark, A.D., Michon, F., Ji, D., Penagos, H., Buzsáki, G., Wilson, M.A., Kloosterman, F. and Chen, Z., 2018. Real-time readout

of large-scale unsorted neural ensemble place codes. *Cell Reports*, *25*(10), pp.2635–2642.

18. Tu, M., Zhao, R., Adler, A., Gan, W.B. and Chen, Z.S., 2020. Efficient position decoding methods based on fluorescence calcium imaging in the mouse hippocampus. *Neural Computation*, *32*(6), pp.1144–1167.

19. Ghosh, K.K., Burns, L.D., Cocker, E.D., Nimmerjahn, A., Ziv, Y., El Gamal, A. and Schnitzer, M.J., 2011. Miniaturized integration of a fluorescence microscope. *Nature Methods*, *8*(10), p.871.

20. Strange, B.A., Witter, M.P., Lein, E.S. and Moser, E.I., 2014. Functional organization of the hippocampal longitudinal axis. *Nature Reviews Neuroscience*, *15*(10), pp.655–669.

21. Lavenex, P. and Amaral, D.G., 2000. Hippocampal-neocortical interaction: A hierarchy of associativity. *Hippocampus*, *10*(4), pp.420–430.

22. Ranganath, C. and Ritchey, M., 2012. Two cortical systems for memory-guided behaviour. *Nature Reviews Neuroscience*, *13*(10), pp.713–726.

23. Haggerty, D.C. and Ji, D., 2015. Activities of visual cortical and hippocampal neurons co-fluctuate in freely moving rats during spatial behavior. *Elife*, *4*, p.e08902.

24. Munoz-Lopez, M., MohedanoMoriano, A. and Insausti, R., 2010. Anatomical pathways for auditory memory in primates. *Frontiers in Neuroanatomy*, *4*, p.129.

25. Jain, L.C., Seera, M., Lim, C.P. and Balasubramaniam, P., 2014. A review of online learning in supervised neural networks. *Neural Computing and Applications*, *25*(3), pp.491–509.

26. Laskov, P., Gehl, C., Krüger, S., Müller, K.R., Bennett, K.P. and Parrado-Hernández, E., 2006. Incremental support vector learning: analysis, implementation and applications. *Journal of Machine Learning Research*, *7*(9).

27. Rezaei, M.R., Gillespie, A.K., Guidera, J.A., Nazari, B., Sadri, S., Frank, L.M., Eden, U.T. and Yousefi, A., 2018, July. A comparison study of point-process filter and deep learning performance in estimating rat position using an ensemble of place cells. In *2018 40th Annual International Conference of the IEEE Engineering in Medicine and Biology Society (EMBC)* (pp. 4732–4735). New York: IEEE.

28. Tampuu, A., Matiisen, T., Ólafsdóttir, H.F., Barry, C. and Vicente, R., 2019. Efficient neural decoding of self-location with a deep recurrent network. *PLoS Computational Biology*, *15*(2), p.e1006822.

29. Cheng, M., Xu, Q., Jianming, L.V., Liu, W., Li, Q. and Wang, J., 2016, November. MS-LSTM: A multi-scale LSTM model for BGP anomaly detection. In *2016 IEEE 24th International Conference on Network Protocols (ICNP)* (pp. 1–6). New York: IEEE.

30. MacDonald CJ, Lepage KQ, Eden UT & Eichenbaum H (2011). Hippocampal "time cells" bridge the gap in memory for discontiguous events. *Neuron* **71**, 737–749.

31. Clancy, K.B., Koralek, A.C., Costa, R.M., Feldman, D.E. and Carmena, J.M., 2014. Volitional modulation of optically recorded calcium signals during neuroprosthetic learning. *Nature Neuroscience*, *17*(6), pp.807–809.

32. Trautmann, E.M., O'Shea, D.J., Sun, X., Marshel, J.H., Crow, A., Hsueh, B., Vesuna, S., Cofer, L., Bohner, G., Allen, W. and Kauvar, I., 2021. Dendritic calcium signals in rhesus macaque motor cortex drive an optical brain-computer interface. *Nature Communications*, *12*(1), pp.1–20.

33. Koester, H.J. and Sakmann, B., 2000. Calcium dynamics associated with action potentials in single nerve terminals of pyramidal cells in layer 2/3 of the young rat neocortex. *The Journal of Physiology*, *529*(3), pp.625–646.

34. Chen, T.W., Wardill, T.J., Sun, Y., Pulver, S.R., Renninger, S.L., Baohan, A., Schreiter, E.R., Kerr, R.A., Orger, M.B., Jayaraman, V. and Looger, L.L., 2013. Ultrasensitive fluorescent proteins for imaging neuronal activity. *Nature*, *499*(7458), pp.295–300.

35. Chen, Z., Blair, G.J., Guo, C., Zhou, J., Romero-Sosa, J.L., Izquierdo, A., Golshani, P., Cong, J., Aharoni, D. and Blair, H.T., 2023. A hardware system for real-time decoding of in vivo calcium imaging data. *Elife*, *12*, p.e78344.

36. Geisler, C., Robbe, D., Zugaro, M., Sirota, A. and Buzsáki, G., 2007. Hippocampal place cell assemblies are speed-controlled oscillators. *Proceedings of the National Academy of Sciences*, *104*(19), pp.8149–8154.

37. Barbera, G., Liang, B., Zhang, L., Li, Y. and Lin, D.T., 2019. A wireless miniScope for deep brain imaging in freely moving mice. *Journal of Neuroscience Methods*, *323*, pp.56–60.

38. Sun, D., Unnithan, R.R. and French, C., 2021. Scopolamine impairs spatial information recorded with "miniscope" calcium imaging in hippocampal place cells. *Frontiers in Neuroscience*, *15*.

39. Ravassard, P., Kees, A., Willers, B., Ho, D., Aharoni, D.A., Cushman, J., et al. (2013). Multisensory control of hippocampal spatiotemporal selectivity. *Science* 340, 1342–1346. doi: 10.1126/science.1232655.

40. Rubin, A., Geva, N., Sheintuch, L. and Ziv, Y., 2015. Hippocampal ensemble dynamics timestamp events in long-term memory. *Elife*, *4*, p.e12247.

41. Pedregosa, F., Varoquaux, G., Gramfort, A., Michel, V., Thirion, B., Grisel, O., Blondel, M., Prettenhofer, P., Weiss, R., Dubourg, V. and Vanderplas, J., 2011. Scikit-learn: machine learning in Python. *The Journal of Machine Learning Research*, *12*, pp.2825–2830.

42. Abadi, M., Barham, P., Chen, J., Chen, Z., Davis, A., Dean, J., Devin, M., Ghemawat, S., Irving, G., Isard, M. and Kudlur, M., 2016. {TensorFlow}: a system for {Large-Scale} machine learning. In *12th USENIX Symposium on Operating Systems Design and Implementation (OSDI 16)* (pp. 265–283).

43. Guizar-Sicairos, M., Thurman, S.T. and Fienup, J.R., 2008. Efficient subpixel image registration algorithms. *Optics Letters*, *33*(2), pp.156–158.

44. Zhou, P., Resendez, S.L., Rodriguez-Romaguera, J., Jimenez, J.C., Neufeld, S.Q., Giovannucci, A., et al. (2018). Efficient and accurate extraction of in vivo calcium signals from microendoscopic video data. *Elife* 7.

45. Lewis, J. P.,1995. Fast normalized cross-correlation. *Vision Interface*, pp. 120–123.

Conclusions and Future Work

6

6.1 KEY FINDINGS

This book uses the hippocampus as an example to demonstrate how an optical BCI, specifically a miniscope, can be applied in both non-real-time and real-time situations. The opening chapter introduces the hippocampus, a crucial brain area linked to memory and cognition. It outlines the evolution of *in vivo* brain signal recording methods, emphasizing the advantages of the recently developed miniscope. In the following chapter, a comprehensive explanation of the optical manipulation process is provided. Readers should gain an understanding of how to utilize the miniscope for recording neuronal signals. The third chapter presents a detailed case study employing the miniscope in an investigation of animal memory. This study characterizes large neuronal populations and explore information processing mechanisms during cognitive impairment induced by scopolamine. Readers should be able to learn basic signal processing techniques commonly used for analysing calcium signals. Chapter 4 illustrates the application of the miniscope in a cognition-related study, demonstrating its capability to detect multisensory modalities in the hippocampus. Readers should become familiar with more advanced methods of analysing neuronal population activity. The fifth chapter showcases a real-time optical BCI application example, utilizing the miniscope to decode spatial and non-spatial information from the hippocampus in real-time. Readers should learn how to decode brain activity in real-time using the raw calcium signal. This book serves as a valuable resource for researchers and neuroscientists, providing insights and guidance on the use of miniaturized fluorescence microscope recording techniques, as well as related advanced analysis methods.

DOI:10.1201/9781003470397-6

6.2 FUTURE WORK

The hippocampal pyramidal cells and interneurons precisely control and modulate neural information processing and communication, and thereupon affects the cognitive functions causally. A considerable amount of research has observed and reported different properties of these two types of neurons, indicating potentially distinct modulation mechanisms[1-2]. Moreover, dysfunctions of hippocampal interneurons have been proposed to impair the hippocampal neural network, leading to cognitive disorders[3]. Hence, it will be critically important to interrogate their activities individually to enhance our understanding of their specific modulation mechanisms. The promoter of the genetically encoded calcium indicator GCaMP6f used in our experiment is synapsin – a protein that expresses in both hippocampal pyramidal cells and interneurons. Thus, the calcium indicator marks all hippocampal neurons indiscriminately, making it difficult for distinction. Remarkably, Dimidschstein et al. (2016) invented a new genetically modified adeno-associated virus (AAV) to express mCherry fluorophores in GABAergic interneurons under the control of mDlx enhancer[4]. This virus can specifically mark the hippocampal interneurons, making it possible to investigate properties of these two types of neurons separately.

Here, a new miniscope recording system prototype was invented and tested on mice to observe two different colour fluorophores. Simply, we continuously injected two viruses (GCaMP6f and mCherry AAVs) in the mouse hippocampal CA1 region using the similar method described in Chapter 2, but there was 15-minutes time interval between two injections. All surgery protocols and postoperation were the same. The new version miniscope (Figure 6.1a) was designed based on the old version and key modifications were: (i) the new miniscope body structure was redesigned to hold two small stimulation LEDs (centre frequency: 568 and 467 nm) with a sliding rail to choose the incoming light source; (ii) the single-band excitation filter, dichroic mirror, and emission filter set were changed to be a dual-band set (# 59022, GFP/mCherry set, Chroma) with custom size cutting; (iii) the system was optimized for focusing. An example of the raw image was shown in Figure 6.1b–c. In this case, the basic recording and processing pipeline should be: (i) detect the spatial location of interneurons (marked with mCherry); (ii) swap the stimulation light source and record the calcium activity of both pyramidal cells and interneurons (marked with GCaMP6f); (iii) deconvolve the raw calcium signal to get spike trains of pyramidal cells and interneurons separately.

In summary, this miniscope prototype makes it possible to observe two different types of neurons separately, which can help gain a more comprehensive understanding of hippocampal pathophysiological dynamics in cognitive disorders. Nevertheless, several questions deserve further investigation:

(a)

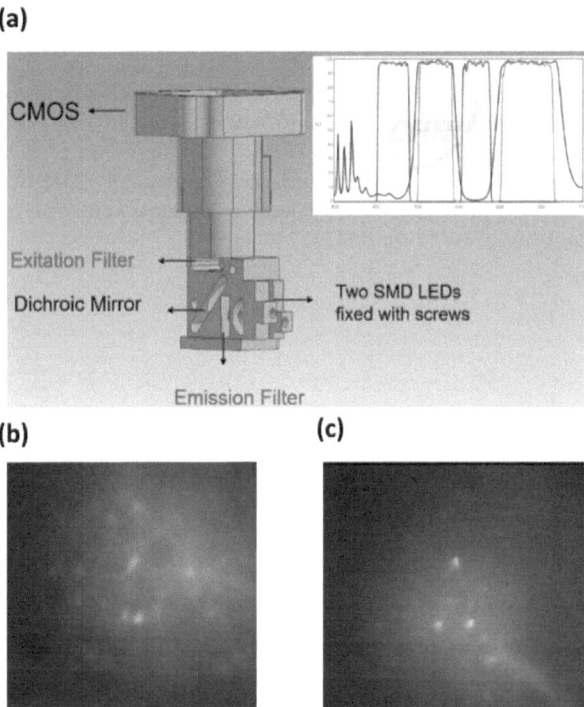

(b)   **(c)**

FIGURE 6.1 The structure of the modified miniscope. (a) Schematic illustration of the new miniscope. (b) An example of the raw calcium signal of hippocampal neurons marked with gcamp6f. (c) The spatial footprints of hippocampal interneurons.

i. What are the differences between the calcium activity of hippocampal pyramidal cells and interneurons in cognitive-related events?
ii. Is it possible to discriminate pyramidal cells and interneurons just based on the raw calcium activity without injecting mCherry AAV that marks interneurons specifically?
iii. How does cognitive impairment affect the calcium activity of hippocampal interneurons?

REFERENCES

1. Klausberger, T. and Somogyi, P., 2008. Neuronal diversity and temporal dynamics: the unity of hippocampal circuit operations. *Science*, 321(5885), pp.53–57.

2. Deng, X., Liu, D.F., Kay, K., Frank, L.M. and Eden, U.T., 2015. Clusterless decoding of position from multiunit activity using a marked point process filter. *Neural Computation*, 27(7), pp.1438–1460.
3. Marín, O., 2012. Interneuron dysfunction in psychiatric disorders. *Nature Reviews Neuroscience*, *13*(2), pp.107–120.
4. Dimidschstein, J., Chen, Q., Tremblay, R., Rogers, S.L., Saldi, G.A., Guo, L., Xu, Q., Liu, R., Lu, C., Chu, J. and Grimley, J.S., 2016. A viral strategy for targeting and manipulating interneurons across vertebrate species. *Nature Neuroscience*, *19*(12), pp.1743–1749.

Index

A

Acetylcholine, 18
Achromatic lens, 88
Adeno-associated virus, 100
Alzheimer's disease, 4, 39, 57
Artificial cerebrospinal fluid, 15, 60
Artificial vowels, 5
Auditory stimuli identification, 74, 82, 94
Autoencoder model, 22
Avalanche duration, 67
Avalanche size, 67

B

Baseplate mounting, 15–16
BCI applications, 57, 88
Branching ratio, 45

C

Calcium activity decoder, 90
Calcium conductance, 33
Calcium dynamics, 95
Calcium kinetics, 7
Calcium transients, 34
Carprofen, 16, 19, 60, 90
Categorization decision, 57
CHEndoscope, 8
CMOS sensor, 7–8, 88
Coding complexity, 41, 56, 61
Cognitive disorders, 4, 100
Cognitive impairment, 2, 99, 101
Cognitive map theory, 39, 59
Constrained non-negative matrix factorization, 21, 61
Convolutional neural network, 82–83
Cosine similarity index, 41–42, 46, 49, 52–53, 56–57, 62
Critical dynamical behaviour, 40, 45
Criticality parameters of neuronal populations, 47
Cross-correlation analysis, 83
Cross-validation, 91
Cyanoacrylate, 16, 19, 60, 88

D

Data acquisition board, 7, 75
Decoding error ratio, 28, 80, 82 83
Decoding noise, 74, 92, 94
Dental acrylic, 60, 88
Deviation from criticality coefficient, 45
Dexamethasone, 16, 19, 60, 90
Diagonalization feature, 22, 26, 30
Dichroic mirror, 7, 88, 100
Dimensionality of the population activity, 44
Diverse cognitive map, 85
Dual-band set, 100

E

Electroencephalography, 2
Electrophysiological recordings, 34
Emission filter, 7, 88, 100
End-to-end analysis pipeline, 84
Enrofloxacin water, 16, 19, 60, 90
Entorhinal cortex, 5, 23, 57
Epilepsy, 3, 58
Episodic memory, 4–5, 18
Euclidean angle, 66
Excitation filter, 7, 88–89, 100
Extracellular calcium, 6

F

Field of view, 2, 7, 77, 88–90, 93
Finchscope, 7
Fine wires, 9
Firing rate map similarity analysis, 95
Flexible electrodes, 9
Functional connectivity, 43, 57, 80

G

Gaussian naive bayes, 76–77
GABAergic interneurons, 100
Genetically encoded calcium indicators, 6
Geometry, 44, 47
Graph theory, 40–43, 57
    clustering coefficient, 43, 48, 50, 54, 65

eigenvector centrality, 65
graph density, 44, 51, 53, 65
GRIN Lens implantation, 15, 19
Gustatory stimuli, 5

H
Half ball lens, 88
Heterogeneity, 46
High channel count BCIs, 9
High-dimensional space, 63, 73
Hippocampal dysfunction, 4
Hippocampal hyperactivity, 3
Hippocampal inhibitory interneurons, 4
Hippocampal network model, 24, 30, 33
Hippocampal population activity, 59
Hippocampal pyramidal cells, 4, 34, 100–101
Hyperparameter optimization, 63
Hyperplanes, 91

I
Image registration, 21, 76, 93–94
Impaired stability, 18, 33
Independent component analysis, 6
Individual neurons, 6, 43, 62
Informational characteristics, 41
Information flow pathways, 57
Inscopix miniscope, 8
Integrative stages, 56, 85
Intracranial electrodes, 6

J
Joint activities, 5

K
Kalman filter, 74, 76, 92, 94

L
Local field potentials, 6, 18, 86
Long-Evans rats, 87
Long short-term memory decoders, 77
Long-term memory, 4
Low gamma oscillation, 4, 57
Lowpass filter, 95

M
Macaque, 86
Manifold amplitudes, 44, 50, 54, 66
Manifold analysis, 39–41, 43, 50, 66
Maximum-likelihood estimation method, 67, 87
Medial septum relay, 56

Memory capacity, 66
Miniaturized fluorescence microscope, 3, 7, 14, 40, 99
Mixed stimuli experiments, 41, 55–56, 58, 63
Modulation mechanisms, 100
Motor cortex, 9
Multilayer perceptron, 77–79
Multi-neuronal microscopy, 7
Multisensory encoding, 39
Multisensory modalities, 3, 99
Multisynaptic pathways, 56
Muscarinic acetylcholine receptors, 18
Muscarinic agonists, 33–34
Muscarinic antagonists, 33
Muscarinic blockade, 32, 34

N
Neural information processing, 5, 100
Neural manifolds representation, 45, 49, 52, 56
Neural oscillations, 3
Neural state space, 57, 66
Neurodegenerative disease, 58
Neuronal avalanches, 45, 47, 67
Neuronal firing pattern extraction, 62, 66
Neuronal sensitivity, 64, 90
Neuron centroid detection, 94
NINscope, 7
Noise level, 76, 79, 83, 95
Non-spatial memory, 5
Non-spatial stimuli, 39, 47, 50–51, 74, 80
Normalized cross-correlation, 95
Normalized fluorescent intensity, 91

O
Odours, 5
Optical brain–computer interfaces, 17, 86

P
Paralysed persons, 9
Pathophysiological dynamics, 100
Phase precession, 32
Photobleaching, 58–59
Place cell activity, 87
Population vector overlap, 22
Position sensitive neurons, 94
Power–law distributions, 45, 47, 52, 56, 58, 67
Precise coding, 42, 90
Principal component analysis, 6
Probabilistic-based prediction, 91
Psychiatry, 58

Q
Quantum efficiency, 7

R
Raspberry Pi, 40, 61, 75, 94
Real-time brain–computer interface, 8
Real-time position reconstruction, 93
Rectified linear unit, 91
Relational theory, 39, 41, 55, 59
Repeated stimuli, 40
Rhythmic light flicker, 4, 57
Robotic arms, 9
Running speed, 21, 25, 29, 32, 87
Running trajectories, 76, 85

S
Scaled exponential linear unit, 22
Schizophrenia, 3
Shape collapse error, 45
Shuffling method, 5, 90, 94
Silicone adhesive, 16, 88
Silicon probe, 6
Similarity between neural modes,
    44, 66
Single-photon imaging technique, 74
Softmax activation function, 92
Spatial cognition, 4, 57
Spatial information content, 21, 25, 76
Spatial memory encoding, 18
Spatial memory retrieval, 18
Spatial navigation, 25, 34
Spatial tuning, 32
Spike duration, 33
Spike sorting, 6, 9, 33
Stimulation LED, 16, 88, 100

Stimuli-dependent topology, 43
Support vector machine, 77
Surface electrodes, 6
System criticality, 58

T
Task-related experiments, 4–5
Temporal coding processes, 46
Temporal spike activity, 21, 23
Temporal stability, 32, 42, 45, 52
Tensorflow, 22, 62, 91
Tetrodes, 6
Time-series data, 86, 92
Time window length, 23
Tones, 61, 74–75, 82, 94
Two different colour fluorophores, 100
Two-photon calcium imaging, 9

U
UCLA miniscope, 7, 14, 59
Utah Array, 9

V
Visual cortex, 38, 85
Visual stimuli identification, 74, 77, 85
Virus infusion, 15, 19
Volitional movement, 9

W
Weighted normalized mutual information,
    42, 63
Widefield imaging, 54

Z
Zemax simulation, 89